D0458323

NOW

NOW

THE PHYSICS OF TIME

Richard A. Muller

W. W. NORTON & COMPANY
INDEPENDENT PUBLISHERS SINCE 1923
New York | London

Copyright © 2016 by Richard A. Muller

All rights reserved
Printed in the United States of America
First Edition

For information about permission to reproduce selections from this book,
write to Permissions, W. W. Norton & Company, Inc.,
500 Fifth Avenue, New York, NY 10110

For information about special discounts for bulk purchases, please contact
W. W. Norton Special Sales at specialsales@wwnorton.com or 800-233-4830

Manufacturing by LSC Harrisonburg
Book design by Chris Welch
Production manager: Louise Mattarelliano

Library of Congress Cataloging-in-Publication Data

Names: Muller, R. (Richard), author.
Title: Now : the physics of time / Richard A. Muller.
Description: First edition. | New York ; London : W.W. Norton & Company,
[2016] | ?2016 | Includes index.
Identifiers: LCCN 2016012496| ISBN 9780393285239 (hardcover) |
ISBN 0393285235 (hardcover)
Subjects: LCSH: Space and time. | Entropy. | Physics—Philosophy.
Classification: LCC QC173.59.S65 M85 2016 | DDC 530.11—dc23 LC record
available at http://lccn.loc.gov/2016012496

W. W. Norton & Company, Inc.
500 Fifth Avenue, New York, N.Y. 10110
www.wwnorton.com

W. W. Norton & Company Ltd.
15 Carlisle Street, London W1D 3BS

4 5 6 7 8 9 0

CONTENTS

Introduction 7

PART I. AMAZING TIME

PART II. BROKEN ARROW

INTRODUCTION

Now—that enigmatic and ephemeral moment that changes its meaning every instant—has confounded priests, philosophers, and physicists, and with good reason. Understanding *now* requires knowledge of relativity, entropy, quantum physics, antimatter, backward time travel, entanglement, the Big Bang, and dark energy. Only *now* do we have all the physics in hand to understand *now*.

The elusive meaning of *now* has been a stumbling block in the development of physics. We understand time dilation from velocity and gravity, even the flipping of time in relativity theory, yet we've made no progress in explaining the most striking aspects of time: its flow and the meaning of *now*. The basic drawing board of physics known as a spacetime diagram ignores these issues, and physicists sometimes perversely treat this absence as a strength and conclude that the flow of time is an illusion. That's backward. As long as the meaning of *now* eludes us, further advances in understanding time—that key aspect of reality—will continue to be stalled.

My goal in this book is to bring together the essential physics, assembling pieces like a jigsaw puzzle until a clear picture of *now* emerges. For this process to work, we also have to find and remove jigsaw pieces that have been mistakenly jammed into the wrong places.

The broad range of relevant physics explains why the puzzle has remained elusive. Physics is not simple and linear, and by necessity this book covers an enormity of material, perhaps too much for a single volume. Feel free to skip around, using the index to go back to key ideas you may have missed. This story can also be thought of as a mystery, with a gradual accumulation of clues leading to a remarkable resolution.

My background is primarily experimental physics, the building and using of new hardware to measure and occasionally discover physical truths that were previously hidden. Two of my projects were directly related to our understanding of time: a measurement of the microwave debris from the Big Bang, and a precise determination of the past expansion of the universe, including the discovery of the dark energy that is accelerating that expansion. Although I admit to having written some purely theoretical papers, I did so mostly when funds to perform experiments were low, or when I thought theory was far off track. As far as I know, this is the only current book written specifically about time by a physicist deeply involved in experimental work, and I will try to give some insights on the challenges and frustrations that such work entails.

The path toward the understanding of *now* requires five parts.

In Part I, *Amazing Time*, I begin by discussing some solidly established yet still surprising aspects of time, uncovered principally by Albert Einstein. Not only does time stretch, flex, and flip, but such behavior affects our daily lives. GPS, the satellite system that keeps us from getting lost, depends exquisitely on Einstein's relativity equations, on these strange properties of time. It is relativity that brought us to think of four-dimensional space-time. The most important message of Part I is that we do understand much about time, and that the behavior of time is not simple but is well established. The pace of time depends on local conditions of velocity and gravity, and even the order of events— which event came first—is not a universal truth. Moreover, Einstein's theory of relativity gives us much of the structure that we need to understand the meaning of *now*.

In Part II, *Broken Arrow*, I remove one jigsaw piece that has been crammed in the wrong spot, the theory that has most inhibited progress in understanding *now*. This improperly placed piece of the puzzle is the theory of physicist Arthur Eddington, purporting to give an explanation for the arrow of time, the fact that the past determines the future and not the other way around. I do this by first presenting the best possible case in support of his theory, and only afterward showing its fatal flaws.

Eddington attributed the flow of time to the increase in *entropy*, a measure of disorder in the universe. We now know enormously more about the entropy of the universe than did Eddington in 1928 when he proposed the theory, and I'll argue that Eddington got it backward. The flow of time causes entropy to increase, not the other way around. Entropy production does not exert the tyranny that is often attributed to it. Control over the pathways of entropy turns out to be essential for our understanding of *now*.

Part III, *Spooky Physics*, brings in another important element for the understanding of *now*: the mysterious science of quantum physics. Quantum physics is perhaps the most successful theory of all time—with agreement between predictions and observations reaching ten decimal places—yet this theory is both disconcerting and distressing. The wraithlike behavior of quantum waves and their measurement blatantly violate Einstein's principles of relativity, but not in any way that can be directly observed or exploited. This behavior of the quantum wave challenges and develops our sense of reality, a sense that will prove essential for the elucidation of *now*. A most disturbing—or perhaps liberating—consequence of quantum physics is that the past no longer determines the future, at least not completely. Some of the most non-intuitive aspects of quantum physics, particularly a strange feature called *entanglement*, have been experimentally verified, and that (surprising!) experimental result suggests that limited ability to predict the future will remain a fundamental weakness of physics forever.

In Part IV, *Physics and Reality*, I explore the limitations of physics. Don't worry—time and *now* do not fall in this realm; they originate in physics, but our perception of them depends on our sense of reality, a sense that stretches beyond physics. Math represents a world of reality that cannot be verified by physics experiment, even something as simple as the irrationality of the square root of 2. But there are other issues that are real but not in the realm of physics, questions such as, what does the color blue *look like*? The denial of nonphysics, nonmath truths has been named *physicalism* by philosophers. Physicalism is faith based and has all the trappings of a religion itself. Alas, against Einstein's fervent hopes, the evidence leads to the conclusion that physics is incomplete, that it never will be capable of describing all of reality.

In Part V, *Now*, the clues fit together to complete the puzzle and provide a unified picture of the reason time flows and the meaning of that ephemeral moment we call *now*. The solution is found in a 4D Big Bang approach. The explosion of the universe continuously creates not only new space but also new time. The forefront expanding edge of time is what we refer to as *now*, and the flow of time is the continual creation of new *nows*. We experience the new moment differently from the preceding ones because it is the only one in which we can exercise choice, our free will, to affect and alter the future. Despite arguments from classical philosophers, we now know that free will is compatible with physics; those who argue otherwise are making a case based on the religion of physicalism. We can influence the future using not only scientific but also nonphysics knowledge (empathy, virtue, ethics, fairness, justice) to guide the flow of entropy to bring about a strengthening of civilization, or its destruction.

I explore three possible tests for this 4D model of progressing time. The observed acceleration of the expansion of the universe, related to the *dark energy*, should be accompanied by an acceleration of the rate of time. This theory predicts that current time is flowing more rapidly than past time did, which leads to the prediction of a new and (possibly) observable time dilation, a new redshift. Effects might also be seen in our study of the early moments of the Big Bang, the era of *inflation*, an epoch we hope to examine by the detection of gravitational waves that were emitted back then, waves we can observe indirectly by studying patterns of polarization in the microwave radiation.

The third test was conceived when the LIGO (Laser Interferometer Gravitational-Wave Observatory) reported in 2016 an amazing detection of two large black holes merging. Such events create new space, and according to the 4D theory, also new time, time which should cause a delay in the latter part of the pulse that could be observed if future events are larger or closer and are detected with a stronger signal.

For those who want more math, details of relativity theory and mathematical results appear in a set of appendices, along with some fanciful poetry and thoughts about nonphysics reality.

Let's start assembling the jigsaw puzzle.

PART I

AMAZING TIME

1

The Entangled Enigma

*Great philosophers were desperately confused about time—
but physics has established hope for understanding it.*

> Time flies like the wind—
> Fruit flies like bananas.
> — *Children's pun*

Here is a fact about you, one that very few people know—maybe no one other than you yourself: you are reading this book right now. In fact, I can be more precise: you are reading the word *now* right now.

Moreover, I stated something that you knew was true, but that I personally didn't know and still don't know. You are reading the word *now* right now, yet I am completely oblivious of that fact—unless, of course, I am standing over your shoulder and you are pointing your finger at the words as you read.

Now is an extremely simple yet fascinating and mysterious concept. You know what it means, yet you'll find it difficult to define without being circular. "*Now* is the moment in time that separates the past from the future." Okay, but try defining past and future without using the word *now*. And what you mean by past and future is constantly changing. A short time ago, reading this paragraph was in the future. Now, most of it is already in the past.

Now that entire paragraph is in the past (unless you are skipping ahead). *Now* refers to a specific time. But the time it refers to is incessantly changing. That's why we use clocks. They report the number associated with *now*; it's called the present time. Clocks update constantly, typically every second. The advance of time is relentless. We can stand still in space but not in time. We move in time but have no control over that movement—unless, of course, time travel proves possible.

The meaning of *now* is just one of the many mysteries of that strange phenomenon we call time. It is remarkable that we understand as much as we do about time, particularly the weird and counterintuitive aspects related to Einstein's theory of relativity, but it is equally remarkable that we understand so little about the fundamentals of time—what it is and how it relates to reality. This book is about time—what we do know and don't know.

Does time *flow*? On April 18, 1906, at 5:12 a.m., a great earthquake shook San Francisco. The time of that event doesn't move; you can look it up in Wikipedia. What does move, what does flow, is the meaning of *now*. *Now* is progressing, changing, advancing through time.

Or maybe it is more meaningful to say that time is flowing past *now*. The whole issue of "movement" is tricky to describe. When we say that an automobile is moving, we note its position at one time and then its position at another time. The velocity is the distance it moved divided by the time it took—given, for example, in miles per hour. That approach for describing *now* fails miserably. *Now* is right now; pause a moment and *now* is still right now. Is it moving? Yes, the movement of time is illustrated by the fact that the meaning of *now* keeps changing. At what rate does time move? One second per second.

There is a third view—that new time is being generated every moment, and that this newly created time constitutes the *now*. Are these views philosophical differences or physical differences? Is it a matter of choice, or is there more truth, more meaning in one or the other? That question is part of what this book will explore.

Suppose time stopped. Would you notice? How? Suppose it moved forward in fits and spurts, or at a totally different rate. Could you detect the difference? Not easily, at least not if you accept the depiction of time commonly used in movies, such as in *Dark City* or *Click* or *Interstellar* or *Lara Croft: Tomb Raider*. Human perception of the movement of *now*, the flow of time, seems to be determined by the number of milliseconds it takes to send a signal from the eye or ear or fingertip to the brain and to record and notice and remember. For humans, that's a few tenths of a second; for a fly, a few thousandths of a second. That's why it is hard for a human to catch a fly. To a fly, your threatening hand approaches in slow motion—just as in *Clockstoppers*.

The rate of time is not just a quandary of science fiction. Relativity gives us specific examples, particularly in the *twin paradox*. One twin, traveling near lightspeed, experiences less time than the stay-at-home twin, yet feels no difference; the two twins experience time in the same manner, even though the flow of time for each is quite different. We'll discuss this strange behavior in some detail.

The hope for understanding *now* is based on the enormous progress that was made in physics in the twentieth century. But let's briefly look back at the frustrations of the ancients.

The Indescribable *Now*

Aristotle's *Physics* dominated science from ancient times until the Renaissance. It was the scientific bible of the medieval Catholic Church. Galileo's denial of some claims in this book is what brought him to trial. Aristotle devoted four chapters of his book *Physics* to wrestling with the concepts of time and *now*, and he emerged fully befuddled. He wrote,

> For what is "now" is not a part: a part is a measure of the whole, which must be made up of parts. Time, on the other hand, is not held to be made up of "nows." Again, the "now" which seems to bind the past and the future—does it always remain one and the same or is it always other and other? It is hard to say. If it is always different and different, and if none of the parts in time which are other and other are simultaneous (unless the one contains and the other is contained, as the shorter time is by the longer), and if

the "now" which is not, but formerly was, must have ceased-to-be at some time, the "nows" too cannot be simultaneous with one another, but the prior "now" must always have ceased-to-be.*

Are those thoughts deep, or are they just confused? In trying to be precise about *now*, Aristotle got tangled in his own words. We can find some comfort in the fact that such a revered thinker found this subject so intractable.

Augustine, in his *Confessions*, bemoaned his inability to understand time's flow. He wrote, "What is time? If no one asks me, I know; if I wish to explain, I do not know." That lamentation, written in the fifth century, resonates with us in the twenty-first. Yes, we *know* what time is. So why can't we describe it? What kind of knowledge is this that we have?

Augustine's conundrum derives, in part, from his axiom that God is all-powerful and all-knowing and all-everything. He makes an astonishing additional conceptual jump: that God must also be *timeless*. This remarkable thought sets the stage for modern physics—physics that describes the behavior of objects within time in space-time diagrams that make no reference to the fact that time flows or that a *now* exists.

For humans, Augustine says, there is no past or future, but only three *presents*: "a present of things past, memory; a present of things present, sight; a present of things future, expectation." (Was this an inspiration for Dickens's *A Christmas Carol*?) But his dissatisfaction with this understanding seeps out. He says, "My soul yearns to know this most entangled enigma."

Albert Einstein was troubled by the concept of *now*. Philosopher Rudolf Carnap writes in his *Intellectual Autobiography*,

Einstein said the problem of the Now worried him seriously. He explained that the experience of the Now means something spe-

---

*   Aristotle, *Physics*, trans. R. P. Hardie and R. K. Gaye, provided by the Internet Classics Archive, http://classics.mit.edu//Aristotle/physics.html.

cial for man, something essentially different from the past and the future, but that this important difference does not and cannot occur within physics. That this experience cannot be grasped by science seemed to him a matter of painful but inevitable resignation. So he concluded "that there is something essential about the Now which is just outside the realm of science."

Carnap disagrees with Einstein's conclusion, saying "Since science in principle can say all that can be said, there is no unanswerable question left." But one must be very cautious when disagreeing with Einstein. It is remarkably easy to dismiss his broodings as emotional rather than substantially deeper than your own. Einstein's simple claims should never be taken for simplicity of thought. Philosophers sometimes feel that they reach great depth by inventing ponderous portmanteau expressions such as "chronogeometrical fatalism" (which is the assumption of constant speed of light). In contrast, Einstein had a way of saying things that even a child could understand—a skill that has made him the most quoted scientist of all time.

Some theorists, rather than interpreting the absence of time flow in physics as a deficiency (as Einstein did), take it to be an indication of a deeper truth. For example, Brian Greene in his book *The Fabric of the Cosmos* suggests that relativity "declares ours an egalitarian universe in which every moment is as real as every other." He says that we have a "persistent illusion of past, present, and future"—a perspective reminiscent of Augustine. He concludes that because relativity doesn't discuss the flow of time, such flow must be an illusion, not part of reality. To me this logic is backward. Instead of insisting that theory explain what we observe, this approach implies that observations must be twisted to match the theory.

Atheists mocked Einstein for drifting away from physics and developing a religious faith in his later years. But they never spoke to his concern that science could not address even these most essential aspects of the world: the flow of time and the meaning of *now*. Many scientists assume that something that cannot be probed by physics is not part of reality. Is that statement a testable claim, or a religious belief itself? Philosophers give this dogma the name *physicalism*. Is there a way to test, to prove, a faith that physics encompasses all? Or is such a belief expected for all physicists, just as being Christian has been an informal but effective requirement to qualify as a potential US president? If you challenge physicalism, do you risk being mocked for your drift toward religion, as Einstein was?

Sir Arthur Eddington is revered among physicists for many contributions, both experimental and theoretical, but he is particularly remembered for his apparent breakthrough in explaining the *arrow* of time, the mysterious fact (to those who ponder such things) that we remember the past and not the future. Yet even though he offered an explanation for the direction of time, Eddington was mystified by its flow. In his 1928 book, *The Nature of the Physical World*, he wrote, "The great thing about time is that it goes on." Then he lamented, "But this is an aspect of it which the physicist sometimes seems inclined to neglect."

In his book *A Brief History of Time*, Stephen Hawking doesn't even mention the *now* conundrum. His book is focused on what we *do* understand, and where the current theoretical activities lie. Hawking talks about the arrow of time but not the flow, about the relativity of time but not the mystery of *now*. Virtually all the recent books on time do likewise. They focus more on potential theories to "unify" the equations of physics than on theories to clarify the meaning of *now* and its flow.

But there is hope.

Broken Symmetry

Coming to grips with the concept of *now* sets us on a journey through abstract and amazing physics, the physics of time, the meaning of reality, and an updated examination of free will. We start by discussing the wonderful and weird behavior of time that borders on the unbelievable but is nevertheless solidly established. The greatest breakthroughs date back to the early 1900s, when Einstein discovered that the rate of time depends on both velocity and gravity. Time is flexible, stretchable, and even reversible. These effects are so substantial that they are built into the current GPS satellites. If GPS didn't adjust for Einstein's discoveries, they would mislocate us by miles. Have a cell phone? Then you are carrying a device that uses relativity right in your pocket.

The strangest aspects of time are found in black holes, those mysterious objects that we are now finding all over the cosmos. Fall into a black hole and as you are being sheared to shreds, you will (according to current theory) travel not only to infinity but *beyond*, as we'll discuss. Look at black holes with fresh eyes and you'll see far more than blackness. You don't have to fall into a black hole to have your sense of reality stressed. Black holes are also relevant to the arrow of time; the current theory (not yet verified) is that they (along with an "event horizon" at infinity) hold most of the entropy of the universe.

Then we'll explore the post-relativity world, when Eddington pondered the direction of time and concluded that it was set by one particular law of physics, the *Second Law of Thermodynamics*, a law stating that the level of disorder in the world, measured by its entropy, increases and will continue to increase forever. This is a strange law, built not on a foundation of physics, but on the fact that our universe is peculiarly well organized and the laws of probability say there is no direction to go but down, toward increased confusion and randomness, heading onward ultimately to a *cold death*. Is that our future? Not necessarily. Increasing confusion in the universe is paradoxically accompanied by increasing organization, associated with the formation of planets, life, and civilization.

I'll show that there are serious alternatives to the entropy arrow of time, including some mysterious aspects of quantum physics that are not understood. The "theory of measurement" is frequently quoted and referenced (239 million hits on Google), but in fact, no such theory exists. The most dramatic discovery in measurement was the experimental confirmation of some strange properties of *entangle-*

ment, a phenomenon that requires covert action to travel faster than the speed of light. It is possible that hiding in the yet-to-be-discovered theory of measurement is the answer to some of our quandaries about time. Quantum physics will play a key role in our untangling of the meaning of *now*.

Some think that time is part of our consciousness that will never, can never, be reduced to physics. Although most physicists believe that all of reality is in their dominion, I'll show that it isn't—that there is knowledge that is just as real as the observations of science but that could never have been discovered experimentally, and cannot be confirmed by measurement. A simple example is the fact that the square root of 2 cannot be written as a fraction containing only integers. Another is the knowledge of what the color blue *looks like*.

Is the arrow of time psychological? If time were moving backward, would we notice? The great physicist Richard Feynman showed that we could consider positrons, antimatter particles used as fuel in science fiction spacecraft, and currently used in hospitals for medical diagnosis, to be electrons moving backward in time. Can *now* move backward in time too? Can we?

In the end, I'll argue that the cause of the flow of time, and the meaning of the mysterious and elusive *now*, does lie within the purview of science—not in the concept of entropy, but in the physics of cosmology. To understand *now*, we have to bring together not just relativity and the Big Bang, but an understanding that the onslaught of entropy has limits. We will have to explore the implications that quantum physics brings to this subject, particularly (and perhaps surprisingly) to the meaning of free will. This new understanding of free will, although not needed for the explanation of *now*, will be important for recognizing why *now* is so important to us.

SPACE AND TIME TOGETHER provide the stage on which we live and die; it is the stage upon which classical physics makes predictions. But until the early 1900s, the stage itself wasn't examined. We were supposed to notice the story, the characters, the plot twists, but not the platform. Then, along came Einstein. His great genius was in recognizing that the stage was within the realm of physics, that time and space had surprising properties that could be analyzed and used to make predictions. Even if he despaired of understanding *now*, his work is central to our understanding. Einstein gave physics the gift of time.

2

Einstein's Childhood Regression

The key questions about time are the simplest ones . . .

Truly I tell you, unless you change and become like
little children, you will never understand time.
— *Apologies to Matthew 18:3*

Despite its appearance, the following quote does *not* come from a children's book on telling time:

> If, for instance, I say, "That train arrives here at 7 o'clock," I mean something like this: "The pointing of the small hand of my watch to 7 and the arrival of the train are simultaneous events."

That deceptively elementary sentence appeared in the premier physics journal of its day, *Annalen der Physik*, on June 30, 1905. The article was arguably the most profound and important article in physics since 1687, when Isaac Newton had effectively jump-started the field of physics with his publication of *Principia*. The author was the man who would become the icon of genius, of scientific productivity, the man ninety-five years later designated by *Time* (appropriately named!) magazine as the Man of the Century, an honor that few disputed. Those words about the small hand of his watch were written by Albert Einstein.

Figure 2.1. Albert Einstein, in 1904, the year before relativity.

The title of Einstein's paper was "On the Electrodynamics of Moving Bodies." What do little hands on clocks and the arrival of trains have to do with electrodynamics, the study of electricity and magnetism? A lot, it turns out. Einstein's paper is really about time and space, and his goal was to bring them into the study of physics. A more appropriate title might have been "The Theory of Relativity—A Revolutionary Breakthrough in Our Understanding of Time and Space." Prior to Einstein, time and space were merely the coordinates used to pose a problem and state the solution. "When will the train arrive?" The answer would be given as a time. Einstein showed that it wasn't so simple.

The Theory of Relativity

What is time? It is rather difficult to define. Newton cavalierly avoided the issue. In his monumental work *Principia*, he wrote, "I do not define time, place and motion, as being well-known to all." Well-known maybe, but difficult to pin down. Einstein didn't define time either, but he examined it with remarkable insight, and in doing so he discovered totally unexpected features. Einstein continues his initial relativity paper in the same almost ridiculously elementary and sometimes boringly pedantic style:

> If at the point A of space there is a clock, an observer at A can determine the time values of events in the immediate proximity of A by finding the positions of the hands which are simultaneous with those events.

Who is Einstein addressing? Rank amateurs? Isn't he stating the obvious? Why did he adopt this childlike voice?

He did it with good reason. To make progress, Einstein had to demolish the hidden prejudices and assumptions that were held, unknowingly, by his professional colleagues. To do that, he first had to uncover them, expose them as things that were not necessarily obvious after all, and—more important—not true. He had to go back to the most fundamental first principles—principles that were taught to you as a child when you first learned how to read a clock, principles such as the universality of time, that even if clocks are sometimes inaccurate they can aspire to being synchronized, that when your father says do it *now*, the meaning of now for you and for him is identical.

He had to remove a jigsaw piece that had been mistakenly jammed into the wrong place.

Einstein had concluded that several seemingly obvious, self-evident principles were not true. His reasoning was based on the theory of electricity—hence the title of his paper. The difficulty of his theory of relativity comes not from advanced math—the paper uses only elementary algebra—but from the flawed way that his readers, the top scientists around the world, thought about time and space.

Try to make yourself think about time and space once again as a child does. Can you remember when you thought that the pace of time was not constant? For me, time went fast during summer vacations and anytime I was having fun. It went slowly when I was visiting the dentist (who didn't believe in using anesthetics) or was waiting for my mom to pick out shoes for herself at the department store. According to a 1929 report in the *New York Times*, Einstein himself said, "When you sit with a nice girl for two hours you think it's only a minute, but when you sit on a hot stove for a minute you think it's two hours."

Ten years after his seminal relativity papers, Einstein published an elaboration, an explanation of gravity that he called the *general theory of relativity*. At that time, Einstein decided that his earlier theory, which didn't include gravity, should be renamed the *special theory of relativity*. That was an unfortunate and confusing name change. It would have been clearer if Einstein had called his original work just plain "relativity" and the later theory "extended relativity." He harbored hopes to push relativity even further, to redo the basic theories of electricity and magnetism and to include them in a *unified* theory, but he never succeeded.

Where does the name *relativity* come from? To understand it, pause for a moment and answer the following question: What is your current velocity?

Don't read ahead until you've come up with an answer to that question. Don't bother to second-guess my intentions; they are benign. Just answer the question. What is your current velocity?

Did you say "zero" because you are sitting still? You might say zero even if you are sitting in an airplane flying at 39,000 feet. The seatbelt sign is on, and you've been instructed not to move about. Since you are not moving about, your velocity must be zero.

Or did you say "550 mph," since that is the speed of the airplane? Or maybe you are reading this book on a slow boat at the mouth of the Amazon and you said "1,000 mph," since that is the rotational speed of the Earth at the equator (24,000 miles around in 24 hours). Maybe you know enough astronomy that you included the velocity of the Earth around the sun and said "18 miles per second." If you used the velocity of the sun around the Milky Way, and the velocity of the Milky Way through the universe (defined by the cosmic microwave radiation), you might have answered "a million miles per hour."

Which answer is correct? Of course they all are. Your velocity depends on the platform you use as reference—what physicists call the *reference frame*. The reference frame could be the ground, an airplane, the core of the Earth, the sun, or the cosmos. Or anything in between.

As you fly in the airplane, do you disagree with someone on the ground about your velocity? No, such disagreement is silly. You both know you are at rest with respect to the airplane and moving at 550 mph with respect to the ground. Both answers are correct.

The startling new feature of relativity is that not just velocity, but time itself, depends on the reference frame. The universal time that you learned about from your parents and teachers does not exist. Not only will you get different times depending on the reference frame you pick: ground, airplane, Earth, sun, or cosmos, but you'll get different *rates* of time. That means that the time between two events, two ticks on your watch, is not universal but depends on which frame you choose.

If you've looked at other popular books on relativity, you have probably read that different observers, moving at different velocities, "disagree." That's nonsense. Even though some of the greatest physicists in the world use those words, they know they are not true. (Full disclosure: I, too, fell into that trap in one of my early papers on relativity. I thought it was a helpful way to teach the subject. I was wrong.)

Statements about observers disagreeing have caused more confusion and caused more people to stutter in their study of relativity than has any mathematical difficulty. Observers in relativity disagree only to the extent that they would also disagree on the velocity of someone flying in an airplane. They all know that velocity is relative, and the number depends on the reference frame, and they (if they have studied relativity) know that the same is true for time. The glory of relativity is that *everyone everywhere agrees*.

Perhaps when I asked you for your velocity, you assumed it was a trick question and you refused to answer. You thought to yourself, *with respect to what?* That's fine too. You correctly guessed where I was going.

Slowing Time

Specifically, Einstein showed that the time when an *event* takes place depends on the reference frame: ground, airplane, Earth, sun, or cosmos. The times will be different. For slow velocities (meaning a million miles per hour or less), that difference will be small, but they will be different. When the frames move rapidly—that is, near the speed of light—the times begin to differ by large amounts. The equations for time in different reference frames are not difficult; they are just algebraic formulas involving squares and square roots. I give them in Appendix 1.

Let's consider a numerical example. Suppose you are in a spaceship moving at 97 percent of the speed of light, with respect to the Earth. We begin with time intervals because they have a particularly simple formula. In the spaceship frame, the time interval between your birthday celebrations is one year. In the Earth frame, the time interval between those same two birthdays is not one year but only three months. I'll show you how to do the math in a moment.

This is what a thoughtful observer on Earth would say: "The time interval between the two birthday parties (the two events) was three months in the Earth frame, and one year in the rocket frame." The observer on the rocket would say exactly the same words. Observers don't disagree on time intervals any more than they disagree on velocities.

Which frame are you in—you yourself? That is a trick question. Answer it anyway.

You are in all frames. Frames are just things used for reference; pick any one you want. If you have zero velocity in one of those frames (say, for example, you are sitting still in an airplane), then we call that frame your *proper* frame. In the sun's proper frame (in which it is at rest), you have a velocity of 18 miles per second, moving around it in a year.

You may be confused on this subject if you have read other books about time dilation, with explanations such as "a moving clock *appears* to be ticking more slowly than yours." Yes, but that's not the whole truth.

Not only does it *appear* to be ticking more slowly, but it *is* ticking more slowly—measured in your proper frame. In its own proper frame, it is ticking faster than in yours. This is not a conflict or a disagreement, any more than the velocity of the person in the airplane (0 or 550 mph?) was a conflict. All observers know; all observers agree.

Lightspeed is defined as velocity divided by the speed of light, just as *Mach number* is velocity divided by the speed of sound. Light (in a vacuum) travels at lightspeed 1. Move at half the speed of light and you have lightspeed 0.5. The time dilation factor, the stretching that takes place when you compare time intervals in two reference frames, is called gamma (the Greek letter γ), and its formula is $1/\sqrt{(1 - b^2)}$, where b is the lightspeed.

In a spreadsheet, if B1 is the lightspeed, then gamma is = 1/SQRT(1-B1^2). For the spaceship example, plug in B1 = 0.97 (lightspeed) and you'll find that gamma (the time dilation factor) is approximately 4. That means that one year on the spaceship takes about four years on Earth. Put another way, the time on the spaceship flows roughly one-quarter as fast as the time on Earth. Spend one year on the ship and you will age only three months. It is ironic, perhaps even amazing, that although we have difficulty defining what we mean by the flow of time, we have a precise formula for relative flow rates.

I encourage you to play with the formula, with a spreadsheet or programmable calculator. You'll find that at lightspeed 0, gamma is 1, so no time stretching takes place when you are at rest. If you try lightspeed 1, you'll find that gamma is 1 divided by 0—that is, infinite. That means that when an object moves at the speed of light, its time (in the Earth frame) comes to a halt. A second in the proper frame of the object takes infinite time in the Earth reference frame.

The relativity of time is readily measurable, at least for an experimental physicist. When I was a graduate student at UC Berkeley, time dilation was an everyday experience. I was working with elementary particles called *pions*, *muons*, and *hyperons*, and they were radioactive. (Individual radioactive particles are harmless; it is only when you have billions of them that they can do significant damage.) Radioactive

Figure 2.2. The author, working with a cyclotron at the Lawrence Berkeley Laboratory in 1976.

particles spontaneously "decay"—a better word is "explode"—and on average, for a given kind of particle, they have a 50 percent chance of decaying in one *half-life*.

The half-life of uranium is about 4.5 billion years, for radiocarbon about 5,700 years, and for tritium about 13 years. In my watch I have tritium mixed with phosphor. (Tritium radioactivity is so weak that it doesn't even get out of the watch hands.) It glows at night, but in 13 years, it will be only half as bright. The radioactivity decays with time. (That's why we call the individual explosions "decays.") The pions in my lab have a much smaller half-life, about 26 billionths of a second (26 nanoseconds). That may seem short, but only to humans; to an iPhone it is long. My iPhone's internal clock runs at 1.4 billion cycles every second. It can do thirty-six elementary computations in the 26 nanoseconds it takes the pion to decay.

At the Lawrence Berkeley Laboratory, where I've done most of my experimental work, I was looking at rapidly moving pions, lightspeed 0.9999988. We had made a beam of pions to see what would happen when they collided with protons. The half-life that I measured for them was indeed about 637 times longer than the value for stationary pions; the measured number matched the calculated gamma factor for that velocity. I was a graduate student, and previously for me, relativity had been only an abstract theory taught in lectures and books. Seeing it in real life was dramatic.

In the physics department at Berkeley, we've now set up a laboratory in which undergraduate students (juniors, typically) can measure time dilation as part of their course work, using not pions but muons, particles created by cosmic rays from space. Relativity is real. For many physicists, it is an everyday occurrence.

Does time dilation mean that when I fly fast in an airplane, I live longer? Yes, and the airplane effect was measured in 1971 by Joseph Hafele and Richard Keating. It was a lovely experiment, one I always talk about when teaching relativity to students. As their platform, Hafele and Keating used an ordinary commercial jet aircraft. Their entire budget was about $8,000—cheap—mostly for around-the-world airplane tickets (including a seat for the clock). The results were published in one of the most prestigious scientific journals, *Science* magazine.

Hafele and Keating had to use a pretty exotic clock to see the effect, but they were able to borrow one. At 550 mph, lightspeed is 0.000000821. To get the time expansion factor, gamma, you can plug this value into the formula, but you'll need a fifteen-place calculator. (Excel won't do, but the iPhone app called *Calculator* works. Orient the iPhone horizontally for the scientific calculator mode.) You'll find that on such a plane ride, you live longer by a factor of 1.000000000000337. That's how much each day is lengthened; the extra time (from the 337 part) works out to 29 nanoseconds (billionths of a second) per day.

Twenty-nine nanoseconds may not seem like much, but in that time the computer in my iPhone can do forty-one computations (it steps through forty-one cycles). Hafele and Keating were able to observe this time dilation and verify that relativity theory gave the right number. Of course, prior to their experiment, physicists had already seen time dilation many times at near-lightspeed velocities, as I had seen in my lab. But it was nice to see the effect at ordinary airplane velocities.

The time dilation effect is even greater for GPS satellites, which have orbital speeds of 8,750 mph, equal to 2.4 miles per second. Do the calculation and you'll find that the time dilation in such a satellite makes time run slower by 7,200 nanoseconds per day. The GPS *must* take that into account, since it uses clocks on the satellite to determine position. Radio waves travel about one foot each nanosecond, so if the 7,200 nanoseconds were ignored, the difference would mislocate you by about 7,200 feet, nearly 1.4 miles.

If Einstein had not discovered the correct equations for relativity in 1905, we might have been puzzled by the long lifetime of pions, and even by the inaccuracy of the GPS in the later twentieth century. We would have discovered time dilation experimentally.

Ride in an airplane, or in a satellite, and from this effect you will live longer—according to an Earth frame. But you will not *experience* more time. Time just runs slower when you are moving. Your clock runs slower, but so does your heartbeat and your thinking and your aging. So you won't notice. That's the amazing thing about relativity. It isn't just clocks that run slower; it is everything. That's why we say that it's the pace of time that changes.

Proper Frames

Einstein discovered that if you restrict yourself to reference frames moving at constant velocity, the equations remain simple. These are the equations I give in Appendix 1. Of course, people don't usually move at constant velocity. We define your proper frame as the one that moves along with you, changing its speed when you do. The most important thing about this frame is that it is the one that determines your age, the amount of time you have to live and to think.

If you sit on the ground and then travel in an airplane and then return, your proper frame keeps accelerating. The amount of time you experience, your age, will be given by your watch. That's actually not obvious, but it is the assumption that all physicists use. Technically, it is called the *chronometric hypothesis*. If you want to know how much you will age on a long and complicated journey, with lots of accelerations, simply keep track of the gamma factor, the formula that tells you how much your clock will slow for every velocity you experience.

For an accelerating frame (such as your proper frame), the general formulas for when an event takes place are much more complicated than for constant-velocity reference frames. To avoid these complications, Einstein used a very simple trick. At any instant, your proper frame will be coincident with a constant-velocity reference frame, so do your instant-by-instant calculations in whatever reference frame happens to be identical at that moment. In other words, if you are accelerating, handle the equations by imagining that your acceleration is essentially a continuous jumping of your proper frame from one reference frame to another one that is moving slightly faster. Einstein later used that approach to make calculations for gravity, which he assumed was equivalent to an accelerating reference frame. He called that assumption the *equivalence principle*.

In this book, when I say "reference frame" I will be referring to a frame that does not accelerate—what physicists usually refer to as a "Lorentz frame," named in honor of Hendrik Lorentz, a contemporary of Einstein, who was the first to use this concept. In contrast, a proper frame moves about with you, starting and stopping and running and walking and changing direction and hopping into automobiles and riding along.

Time Travel into the Future

Time dilation gives a straightforward means of time travel into the future. Move at a sufficiently high velocity and your proper time will slow, and in one minute of your time you can be a hundred years in the future. There is no need to freeze your body and hope that future science will figure out a way to unfreeze it; you just leverage velocity. Of course, there are practical details. You need to make sure that in your travels, you don't bump into anything; at high speeds, that would be disastrous. You need to make sure you'll come back to the right place, that the Earth will be where you expect. And there is a catch. Once you are in the future, there is no similar mechanism that will allow you to come back.

Backward time travel might be possible. People have proposed that faster-than-light travel might do it, or sliding down and through something called a wormhole. I'll discuss these two ideas, but there are serious problems with both approaches, and I'll argue that neither will actually succeed.

Einstein derived his equations assuming that the relative velocity of reference frames was less than the speed of light. If it is equal to the speed of light, gamma becomes infinite and the equations are invalid. Can you use the formulas for velocities greater than the speed of light? Not officially, but of course, everybody tries it to see what happens. You wind up getting imaginary mass. That's not necessarily unphysical. We'll talk about that when we discuss the hypothetical faster-than-light particles called tachyons.

3

The Leaping *Now*

*Changing reference frames makes discrete jumps
in the time of distant events.*

> This day and age we're living in
> Gives cause for apprehension
> With speed and new invention
> And things like fourth dimension.
> Yet we get a trifle weary
> With Mr. Einstein's theory . . .
> You must remember this
> A kiss is just a kiss, a sigh is just a sigh.
> The fundamental things apply
> As time goes by.
> — *Excerpt from the song*
> *"As Time Goes By"*
> *(including lyrics skipped*
> *in the movie* Casablanca)

Even if you feel comfortable with time dilation, Einstein's discoveries about *when* and *now* may trigger distress. The term *quantum leap* was originally used to refer to processes in quantum physics. But *quantum* means "discrete, sudden, abrupt." According to relativity, just such an abrupt change takes place for a distant event when you abruptly change your choice of reference frame. The time jump can be very large.

We give an event a name ("my New Year's party"), and we can designate it by its location and time. My New Year's party was on midnight,

December 31, 2015 (or whenever), and its location was my home—given by three dimensions, such as latitude, longitude, and altitude. The time is the *when*. If two events have the same *when*, they are called simultaneous. Your party and your friend's New Year's party were simultaneous. (Recall Einstein's quote at the beginning of the previous chapter, about the hand of the clock and the arrival of the train.) Simple enough. But, if two events are simultaneous in one reference frame—say, that of my house—are they necessarily simultaneous in another reference frame, such as that of a moving airplane? The obvious answer is yes. The correct answer is no.

Unless you've studied Einstein's work, would it ever occur to you that the answer might be no? His true genius was in his ability to ask such a question. Indeed, without giving up the concept of universal simultaneity, Einstein could not have solved the problem of relativity.

In his theory, Einstein showed that if two events occur at different locations, then simultaneous events—let's say both of them occurring *right now*—would not be simultaneous in another reference frame. One event would occur before the other. Which one came first? That depended on the frame. You could have them in either order. This is what I mean when I say that, in relativity theory, time can flip.

Suppose you travel to a distant star. What is happening back on Earth? Implicit in that question is the understood but unstated word *now*: What is happening back on Earth *now*? But as soon as you halt, arrive at that star, change your proper frame from one moving to one stationary on that star, the meaning of the universal *now* in that frame changes, because your proper frame after stopping is matched to a different reference frame. When your proper frame jumps to a different reference frame, so does the time of a distant event. The formula for the time jump turns out to be remarkably simple: $\gamma Dv/c^2$, where γ is the gamma factor, D is the distance to the event, v is the velocity change, and c is the speed of light. I derive this formula in Appendix 1.

Here's an example. Suppose your New Year's Party is located at your home, and mine is located on the moon. These events are simultaneous in the proper frame of my home. Let's look at these same two events in the proper frame of my laboratory pion. The distance D/c is 1.3 light-seconds, the pion velocity from my lab v/c is close to 1, and gamma is the number I calculated earlier, 637. So the time jump is just the product of 1.3 and 637, equal to 828 seconds. That's 14 minutes between the "simultaneous" New Year's parties! Which event comes first depends on whether the pion frame is moving toward the moon or away from it.

Do you find this example to be more disturbing than the longer life-time? Most people do, yet it is just as real. Because it is so difficult to accept, this time jump is at the heart of the most confusing paradoxes of relativity, discussed in the next chapter. It also has important consequences in our search to understand *now*.

Once again, be careful not to think of this time jump as a "dis-agreement between observers," a vernacular used in many popular explanations of relativity. Observers with different proper frames don't have "different conceptions" of reality, as some authors would have you believe. That conclusion is based on an unstated (and incor-rect) assumption that any observer is restricted to describing reality from the perspective of only one frame, the observer's proper frame. If that were the case in ordinary life, I would have to say not that I went to Paris, but that Paris came to me. We are not restricted to proper frames in everyday life, and there is no reason to so restrict ourselves when talking about relativity.

Squeezed Space, Crêpe Protons

Einstein altered both our understanding of time and the way we think about space. In his relativity paper, he concluded that not only does the time between two events depend on the reference frame (ground, air-plane, satellite) but so also does the length of objects.

To talk about length, we once again have to regress to childhood. To measure the length of a bus, we measure the position of one end and the position of the other end, and then take the difference. But suppose the bus is moving. We measure the location of the front of the bus to be right next to us, and a moment later, the back of the bus is right next to us, so we erroneously conclude that the length of the bus is zero. Well, obviously we made a mistake. We have to measure the positions of the front and back *simultaneously.*

Simultaneously? Ah, that's the catch. That concept is relative. Simultaneous in one reference frame is not simultaneous in another reference frame. A direct consequence is the fact that length will be different in different frames. If an object has a length L in its proper frame (moving with it), then the length in a frame moving at relative velocity v (such as the ground) will be, according to Einstein, shorter by the gamma factor. For those interested, I derive this equation in Appendix 1.

This shortening has had several names: *FitzGerald contraction, Lorentz contraction, length contraction.* The multitude of names reflects the fact that it was postulated even before Einstein. George FitzGerald, along with all the other physicists of his era (late 1800s), assumed that all of space was filled with an invisible fluid called *aether* (British spelling; when I was young I confused this with the chemical ether). Aether was the thing that was waving when light waves and radio waves waved. It was what we now call the *vacuum* or *space.* FitzGerald hypothesized that an object moving through the aether would be compressed by the resistance of this fluid, by what he referred to as the aether wind. The new length would be the old length (the one in its proper frame) divided by gamma.

Length contraction is also confused by the poor language used by some writers. They say that a moving stick "appears to be shorter." True, but not the whole truth. The meter stick *is* shorter in our reference frame than in its proper frame. All observers, regardless of their velocity, agree on this. The meter stick appears to be shorter because it *is* shorter.

Length contraction was also something I could detect in my lab, although I couldn't see it as clearly as I could see time dilation. When we hit a proton with a pion, in the frame of the pion the proton wasn't spherical at all. It was like a very thin pancake, with thickness 1/637 times its diameter—more like a crêpe. That shape change seriously affected the way the pion bounced off the proton, and that altered scattering was what I observed.

In the Earth frame, the pion was the shorter of the two particles. So, which was *really* shorter, the pion or the proton? The answer, of course, is both, depending on the reference frame. In the pion proper frame, the proton was moving, and the proton was shorter; in the proton proper frame, the pion was moving, and the pion was shorter. All observers, in all frames, agree to these facts. In relativity, observers never disagree about lengths any more than they disagree about velocities. Velocity is relative; so are time intervals; so is shape.

The Michelson-Morley Experiment

Most popular discussions of relativity theory start with a description of the experiment done by Albert Michelson and Edward Morley in 1887. Yet it is not clear how much the results of the experiment influenced Einstein; he mentions it only in later papers, and it appears to many that his theory of relativity was based primarily on the properties of the Maxwell theory of electromagnetism, and the properties of that theory deduced by Lorentz.

Michelson and Morley made an extremely sensitive measurement of the velocity of light in two perpendicular directions—that of Earth's motion around the sun, and the perpendicular direction—in an attempt to detect the aether wind. They found that the velocity of light in both directions was the same, despite the Earth's motion. They observed less than one-fortieth the difference between velocities that they expected—essentially no velocity difference at all.

Modern experiments have confirmed that the speed of light is constant to an accuracy of better than 0.01 micron per second, independent of the direction of motion of the Earth. In fact, the accuracy is so precise that a better measurement would require a better definition of what we mean by a meter. To address this problem, the speed of light is now defined to be 299,792,458 meters per second exactly, and the length of the meter is officially defined as the distance traveled by light in 1/299,792,458 second. That means you can no longer improve the known value for the speed of light; you can only improve the accuracy of the size of the meter. A useful value to remember is that light travels about 1 foot in one nanosecond (one-billionth of a second), to an accuracy of 1.5 percent.

The constancy of the speed of light is readily accounted for by relativity theory, as I show in Appendix 1. That fact can be turned around. In teaching elementary courses, instructors sometimes derive the relativity equations by starting with the constancy of the speed of light and then showing that the relativity equations are the only equations linear in time and position that yield this result. As a student, I never liked that derivation because I found the assumption of linearity to be artificial. It is not, but it was hard for me as a sophomore in physics to accept the importance of "linearity," so the whole derivation seemed forced.

$$E = mc^2$$

The most famous equation of the twentieth century is Einstein's formula relating energy to mass: $E = mc^2$. It is now so familiar that it is hard to realize how absurd it seemed at the time Einstein first stated it. He published it in his second relativity paper, in September 1905, three months after his first relativity paper.

The equation was patently ridiculous. It says that *any* mass, even such unburnable substances as rock or water, contains enormous energy. The huge value comes from the inclusion of c^2 in the formula. The speed of light, c, is 300 million meters per second. Square it and you get a value of 90,000 million million—in other words, 90 quadrillion. Making it worse, Einstein gave no means of extracting this energy for useful purposes. He simply said it was there. Unless you had some way of getting rid of mass, such energy was useless. At that time, mass was widely assumed to be immutable. It was "conserved"; it could not be created or destroyed. So the equation seemed both absurd and meaningless.

Einstein was saying that, in principle, all energy is *equivalent* to mass. You can think of mass as bundled up energy. When you extract heat energy by burning gasoline and air, the mass of the fumes (mostly carbon dioxide and water vapor) will be slightly lighter than that of the burned gasoline plus air, because of the missing energy (the energy used to move your auto). That energy goes into heating the air and the ground (through friction), and that means the air and the ground will be slightly heavier, since they contain more energy.

The formula $E = mc^2$ assumes physics units (joules, kilograms, meters per second). Let me rewrite that equation in terms of everyday units. A kilogram of mass weighs about 2.2 pounds, and 1 kilowatt-hour (kWh) of energy is the same as 3.6 million joules. So I can rewrite the equation as

$$\text{Energy} = mc^2 = 11 \text{ billion kWh per pound}$$

In the United States, the average cost of electricity is 10 cents per kilowatt-hour. So one pound of anything converted to electric energy would be worth over a billion dollars.

Another way to write the equation is to measure energy in gallons of gasoline equivalent. While we're at it, let's measure mass in terms of the mass of a gallon of gasoline. Then the equation becomes

$$\text{Energy} = mc^2 = 2 \text{ billion gallons equivalent, per gallon of gasoline}$$

This means that the energy in the mass of the gasoline is 2 billion times greater than the energy available from burning it. In the United States, the price of gasoline is bouncing around these days, but for the sake of this example, let's use the number $3 per gallon. Then the total energy in one gallon of gasoline is worth $6 billion, more in Europe.

Did it take courage for Einstein to publish such an apparently ridiculous conclusion? Today, with nuclear power and nuclear bombs, it doesn't seem so far-fetched, but there was very little evidence back in the early 1900s for this tremendous energy content, except for the fact that in a radioactive decay, the energy released was over a million times greater than the chemical energy in that same atom. There had to be a previously unrecognized source of enormous energy, and Einstein found it: mass. Einstein's claim took either enormous daring or a deep conviction that he had indeed uncovered a fundamental truth of mass. It appears to have been the latter.

How did Einstein draw conclusions about energy content from equations about time and space? To him, the method was straightforward. He asked, What would those changes in our understanding of time and space do to the laws of mechanics? Newton had concluded that a particle feeling a force F will accelerate by an amount a given by the equation $F = ma$. We call that Newton's "second law." (His first law, that an object in motion will stay in motion, is just a special case of the second law, for zero force.)

Einstein realized that Newton's equations could not be true in all frames, so he devised new equations that would be. A key conclusion was that moving particles behaved as if they were heavier. In many of the equations, in place of m, the term γm appears, a combination historically called the *relativistic mass*. Energy was $E = \gamma m c2$, leading Einstein to recognize the equivalence of relativistic mass and energy. (Some physicists currently prefer to use the word mass only for rest mass, but then we lose the equivalence of mass and energy; moreover the concept of relativistic mass was widely used by scientists such as Lawrence, and proved conceptually important.)

Think again about the pion in my laboratory. Not only was its time progressing 637 times slower than my time, and not only was it squished into a crêpe 637 times thinner than its diameter, but its mass was 637 times greater than the value listed in the physics particle-mass tables. Moreover, I could easily measure this mass increase by the way the pion plowed through the strong magnetic field we used in our experiment with minimal deflection. Relativity is indeed real, and to me in my lab, I saw the effects every day.

I could also directly observe the transformation of mass into energy. I was using a liquid-hydrogen bubble chamber, a device invented by my mentor Luis Alvarez. This instrument created a trail of tiny bubbles along the path of the moving particle. The most dramatic decays were those of particles known as muons. When a muon underwent its radioactive explosion, its track suddenly vanished and was replaced by a new track of a lighter but much faster electron. The heavy mass of the muon had been converted directly into kinetic energy, energy of motion, of the electron.

In my lab I also frequently observed antimatter. I'll talk more about antimatter later, but what is most interesting for our current discussion is the fact that when antimatter slows and hits ordinary matter, it annihilates—meaning that it turns all of its mass and the mass of the target into energy, typically gamma rays, which then are absorbed and turn into heat. I was seeing mass converted to heat every day. A matter-antimatter mixture contains the highest energy release of any fuel, a thousand times higher than the energy available even in nuclear fusion, a billion times greater than in gasoline. That's why it was the fuel mix used by the starship *Enterprise* in *Star Trek*.

Matter-antimatter annihilation is also used on a daily basis in hospitals for medical imaging. The name most often used for an antimatter electron is *positron*, and that is the origin of the *P* in medical *PET* scans. A PET scan takes advantage of the fact that some radioactive chemicals, such as iodine-121, emit positrons. In the human body, iodine accumulates in the thyroid gland. When it emits its positrons, those particles find nearby electrons and annihilate, releasing gamma rays. A camera can image where those gamma rays originate and produce a photo of the thyroid gland. Such photos have medical value because if part of the gland is not active—that is, not concentrating iodine—it appears as a blank spot in the image.

Some people mistakenly think that Einstein's equation played an important role in the development of the atomic bomb, but that's not true. Prior to Einstein, radioactive explosions had already been known to release huge amounts of energy. That knowledge, plus the possibility of a chain reaction, was all that was needed for the atomic bomb.

Hungarian physicist Leo Szilard was awarded a classified patent for the bomb, based on these concepts, in 1936. (It was a British patent; by then, Szilard had escaped the Nazis, and he resided in London; in 1937 he moved to New York.) In 1939, a letter drafted by Szilard and signed by Einstein convinced President Roosevelt to initiate research that led to the Manhattan Project to build an atomic bomb. What we learned from the Einstein equation was that the huge energy release results in a small decrease in the mass of the fissioning atoms, but that knowledge was not needed or used in the design of the bombs.

Time and Energy and Beauty*

Some people think of *energy* in the same mystical way they think of *time*. The most remarkable feature of energy, its most useful feature, is also not at all obvious: that energy is *conserved*. What does that mean? If energy is conserved and we have no choice in the matter, why do our ecological leaders tell us to conserve energy? What they really mean is that they want us to minimize *entropy*—avoid making too much of it; we will talk about entropy at length in Part II. We don't have to try to conserve energy; it is automatically conserved.

The relationship between time and energy is very deep and was first recognized by Emmy Noether (pronounced NER-ter), a woman whom Einstein referred to as one of the most significant and creative mathematicians of all time. Noether showed that, in ordinary physics, it is not necessary to *postulate* conservation of energy—that for any set of equations (mechanical, electrical, quantum) there is a principle that can be used to *prove* conservation of energy. That principle is *time invariance*.

Simply, time invariance means that the laws of physics don't change with time. In classical physics, $F = ma$, and that equation is as true today as it was yesterday. Noether's theorem showed that if you assume time invariance, there will always be a quantity, calculated from the elements

* Parts of this section were adapted from my book *Energy for Future Presidents: The Science Behind the Headlines* (New York: W. W. Norton, 2012).

of the theory (mass, velocity, position, field, and so on) that will never change. In classical physics, time invariance implies the conservation of Newton's energy, the sum of kinetic plus potential. Noether's method allowed an unambiguous way to define energy for any new set of equations, such as those of relativity. I discuss Emmy Noether and her amazing deductions in more detail in Appendix 2.

The connection between time and energy goes even deeper. As a result of Noether's work, we now understand why time and energy always appear together in quantum physics. As I'll discuss in some detail in

Figure 3.1. Emmy Noether, who discovered the link between time and energy.

Part III, this understanding inspired Richard Feynman to interpret antimatter as ordinary matter moving backward in time.

Profound associations such as this one, relating two concepts (time and energy) that seem completely unrelated, are what physicists think of as the "beauty" of physics. You don't have to agree that such relationships represent beauty. You may find a rainbow or the eyes of a child to be far more thrilling. But at least this example can help you understand what physicists are referring to.

What's So Special about the Speed of Light?

When I was a teenager and read about relativity (in a wonderful book by George Gamow called *One Two Three . . . Infinity*), I wondered, What is so special about light that its speed c appears in the fundamental equations of relativity? Is light more fundamental than, say, an electron?

As I studied physics, first in high school, then in college and graduate school, and then continuing for the rest of my life, I realized that the answer is that light just happens to be the first thing we knew about that had the peculiar property of having zero *rest mass*—the term in our equations that is written as m. (The relativistic mass is γm.) We now know of other such particles. Gravity waves, in their particle form called gravitons, also have zero rest mass. So we could write the equations of relativity without reference to light and instead use gravitons; the term c would then be the velocity of gravitons. There may also be a neutrino that has zero rest mass; call it the massless neutrino. Then we could call c the velocity of the massless neutrino.

Better would have been to call c the "Einstein velocity." It could also be called the "limiting velocity," the maximum velocity that any moving mass could obtain. It would exist regardless of whether there actually were particles (such as photons) that move that fast. According to the theory, all massless particles will travel at the Einstein velocity c. It is the Einstein velocity that is fundamental. Photons, gravitons, and massless neutrinos (if they exist) all move at this velocity. We also think that, in the very early universe, before the so-called Higgs field developed, *all* particles (including electrons and the constituents of protons, called quarks) were massless and moved with the Einstein velocity, the speed of light.

The ironic part about zero rest mass is that you can never bring a photon or other particle with zero rest mass to rest. It will have zero energy (the gamma factor is 1 and m is 0, so $E = \gamma mc^2 = 0$), so the particle will not "exist." If we try to bring light to rest, for example, by having it absorbed on a black surface, it gives off all of its energy, typically by heating the surface, and no light remains.

Black Holes

Black holes are massive objects that you can fall into but can't get back out of. That mysterious property appears to give a direction to time. Moreover, the study of black holes will lead us to some other properties of time that are even stranger than those we've already examined.

The idea of a black hole dates back to 1763, when English scientist John Mitchell realized that the escape velocity of a star could exceed the speed of light. If light couldn't escape, the star would appear black, he reasoned. He even calculated what turned out to be the right equation. His idea didn't attract attention, because at that time, light was already known to be a wave, and most people mistakenly thought that a wave would not be pulled by gravity. Now we know from relativity that since waves carry energy, they also carry mass, and gravity does indeed pull on them.

To make a black hole, something with a very high escape velocity, we must put a large mass in a small volume. Suppose we stuffed the mass of the sun into a sphere 1 kilometer in radius. From freshman physics, we can calculate that the escape velocity would be 500 million meters per second,* or about 1.7 times the speed of light. Light could not escape from the surface. The compressed sun would be totally black.

---

*   Escape velocity is calculated from $GMm/r = \frac{1}{2}mv^2$, where G is the gravitational constant, M is the mass, and r is the radius. So $v = \sqrt{(2GM/r)} = \sqrt{[2 \times (6.7 \times 10^{-11}) \times (2 \times 10^{30})/1,000]}$ $= 5 \times 10^8$ meters per second.

Relativity makes it possible to derive the properties of a black hole in another way, from the relativistic mass increase. The energy it takes to shoot a satellite into space depends on the mass of the satellite, but the faster you send the satellite, the more mass it has (from relativistic mass increase), and thus high speed increases the gravitational pullback. If the mass of the star is high enough, or its radius sufficiently small, then I can never provide enough kinetic energy to overcome the added binding energy.

The physics jargon is this: the kinetic energy (energy of motion) will always be less than the binding energy (potential energy) to such a star. The satellite will fall back regardless of the speed I give it. This happens when the mass M of the star is stuffed into a radius R. Using scientific notation,* the formula is

$$R = 1.5 \times 10^{-27} M$$

This value of R is called the *Schwarzschild radius.*

The mass of the Earth is 6×10^{24} kilograms. Plugging this value into the equation gives a Schwarzschild radius for Earth of about 0.01 meter, or 1 centimeter. I weigh about 190 pounds = 83 kilograms, so I would be a black hole if you could stuff me into a sphere of radius $R = (1.5 \times 10^{-27})$ (83) = 1.3×10^{-25} meter. That's a billion times smaller than the nucleus of an atom.

We think that there actually is a mechanism by which an object several times heavier than the sun can become a black hole. It involves a supernova explosion in which the outer part of the star is blown away while the inner core collapses. Several astronomical objects that are widely believed to be black holes were formed in this way, including Cygnus X-1, a strong X-ray emitter in the constellation Cygnus.

---

* The expression 10^{-27} = 0.000000000000000000000000001. The 1 appears in the twenty-seventh decimal location. It is equal to the inverse of 10 multiplied by itself 27 times. In spreadsheet notation (used in Excel and scientific calculators), the value is represented as 1E–27. In the following use, for the mass of the Earth, 10^{24} is 10 multiplied by itself 24 times; it is 1 followed by 24 zeros. In spreadsheet notation, it is 1E+24.

This object was the subject of a famous (among physicists) 1975 wager between Kip Thorne and Stephen Hawking. Thorne bet that Cygnus X-1 was indeed a black hole; Hawking bet that it was not. Fifteen years later, in 1990, Hawking conceded that he had lost—and fulfilled his wager by giving Thorne the promised subscription to *Penthouse* magazine. Of course, Hawking also benefited from his concession, since much of his research of the prior decade would have proved meaningless if black holes didn't really exist—as Hawking himself pointed out. Ironically, I'll show later in this book that Cygnus X-1 is *not quite* yet a black hole according to relativity theory, although it is very close.

We know of no mechanism to turn the Earth or me into a black hole.

WHAT BOTHERS PEOPLE most about relativity is not the mysterious black hole, but an apparent contradiction that arises from time dilation. The moving person ages less than the stationary person. OK, but isn't all motion relative? Which one is moving and which is stationary? It sounds like both people are younger.

And it turns out that both are indeed younger, in the appropriate reference frames. But then what happens if one turns around and comes back and they meet? They can't both be younger when they are face-to-face. It is easier to understand time if this and other paradoxes are explicitly addressed.

4

Contradictions and Paradoxes

Relativity looks logically inconsistent,
until you look closely and carefully . . .

All truth passes through three stages. First, it is
ridiculed. Second, it is violently opposed. Third,
it is accepted as being self-evident.
>   — *often attributed to*
>   *Arthur Schopenhauer*

A paradox, a paradox,
A most ingenious paradox.
>   — Pirates of Penzance

Einstein's discovery that time slows for moving objects was astonishing. His discovery that the order of events can be relative was troubling. And his further deductions about energy certainly seemed, at the time, incredible. More than anything, Einstein's study of time showed us that the subject of time is full of surprises, and that the conclusions affect not only our understanding of the universe, but also our everyday lives.

Even after you think you have accepted Einstein's results, some of the consequences continue to surprise. Stated in certain ways, these results lead to apparent contradictions that can drive students (and some professors) nuts. The two most famous and perplexing ones are called the *twin paradox* and the *pole-in-the-barn paradox*. I introduce a third one here, the *tachyon murder*.

Relativity theory is completely self-consistent but, especially to a newcomer in the field, it doesn't seem that way. The apparent contradictions and paradoxes are based on simple mistakes, just as in the proof* that 1 = 2. You might think these paradoxes would bother only beginners, but even experts have prejudices and assumptions that they're unaware of. As a result, even many professors get confused when trying to explain these paradoxes to their students.

I'll begin with the easiest of the paradoxes to understand.

The Pole-in-the-Barn Paradox

A farmer has a barn that is 20 feet long with a door at the front. He has a 40-foot pole that he would like to store inside the barn (Figure 4.1, top). He has studied relativity, so he plans to use the length contraction to fit the pole inside. He runs with the pole fast enough to shorten the length of the pole to 20 feet, meaning that gamma is 2 (Figure 4.1, middle). He plans to close the barn door behind him as soon as the pole is inside. That should work.

But as soon as he starts running with the pole, the farmer realizes that in his new proper frame (the running one), the barn is shorter, not the pole, by a factor of gamma = 2; the barn is only 10 feet long. The running frame is also the proper frame of the pole, so the pole has an unchanged length of 40 feet. There is no way to get the 40-foot pole inside the 10-foot barn (Figure 4.1, bottom).

Yet in the proper frame of the barn, the pole fits easily. So, what actually happened? Did the farmer succeed in getting the pole into the barn

---

* Here's a proof that all numbers are equal. Let A = 13 and B = 13; C and D can be any two numbers. Then A = B. Multiply both sides by (C – D) to get A(C – D) = B(C – D). Expand: AC – AD = BC – BD. Rearrange: AC – BC = AD – BD. Factor: C(A – B) = D(A – B). Cancel (A – B) to get C = D. Since C and D were arbitrary, I've proved that all numbers are equal. The mistake was dividing by (A – B). That's illegal, since A – B = 0. A simpler (but more obviously flawed) version of the proof is this: C × 0 = D × 0. Cancel the zeros.

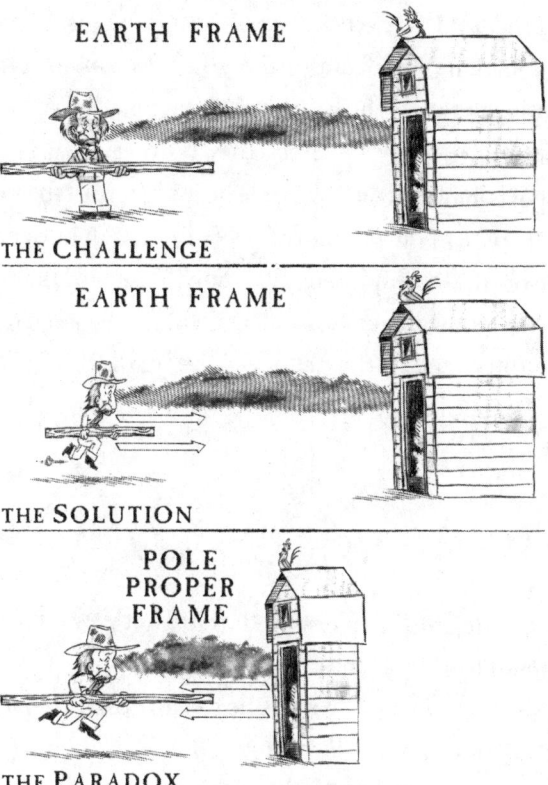

Figure 4.1. The pole-in-the-barn paradox. (Top) The farmer ponders how to get a 40-foot pole into a 20-foot barn. (Middle) He runs fast, and the length of the pole contracts to 20 feet. It will fit! (Bottom) For the same situation described in the farmer's proper (running) frame, the pole is still 40 feet long, but the barn is contracted to 10 feet. The pole clearly will not fit. (Illustration by Joey Manfre.)

or not? How could the answer depend on the reference frame? It's either *in* or *not in*. They can't be both right.

This paradox is easily resolved when worded carefully. By *inside*, we mean that both ends of the pole are in the barn *simultaneously*. This is achieved in the barn proper frame. The front of the pole hits the wall of the barn at the same time that the rear of the pole comes in and the door is closed. But those two events are not simultaneous in the pole proper frame. In that frame, the front of the pole smashes into the wall of the barn, and only later does the back end of the pole enter the door.

As always, both observers agree. They both say that both ends of the pole get inside the barn. In the barn proper frame, the two events are simultaneous. In the pole proper frame, even though both ends get inside, they don't do it at the same time; they don't do it simultaneously. "Being inside the barn" is a statement that cleverly hides the issue of simultaneity.

For those interested in the details of the math, the calculations illustrating this resolution of the pole-in-the-barn paradox are given in Appendix 1.

The Twin Paradox

Consider the twins John and Mary. They are both twenty years old. John stays at home, and Mary goes off on a spaceship at high velocity to a distant planet. Mary's velocity gives her a time dilation factor gamma = 2. According to John, Mary is younger. But according to Mary, John is younger. They can't both be right. What happens if Mary returns? Once they come together, certainly they can tell who is younger. Paradox!

To resolve the paradox, we have to choose our words carefully. We have to watch out for hidden assumptions regarding *simultaneous* that may not be valid, and the implicit assumption that people are required to report their results using only the coordinates in their own proper frame.

On the outward journey, John and Mary completely agree. With reference to John's proper frame, Mary is moving; with reference to Mary's proper frame, John is moving. In John's frame, Mary is younger; in Mary's frame, John is younger.

OK, but what happens when Mary stops, turns around, comes back, meets with John face-to-face, and they compare ages? At that time their proper frames are identical. Which one is younger? They can't both be. And, indeed, they are not.

This paradox has a satisfying resolution, involving the issue of simultaneity. I work out the numbers in Appendix 1, with specific values assumed for the velocities and the distances. Before Mary turns around, in her proper frame John is younger. That means that she is celebrating an older birthday simultaneously with John's younger birthday. But after she turns around, in her proper frame those two events are no longer simultaneous. In the new frame, John is simultaneously celebrating a much older birthday than is Mary.

On Mary's way back, in her proper frame John is the one who is moving, so he is aging less. Nevertheless, the leap in time was so great that when they come together, John is still older than Mary. That's the same result we would get if we did all the calculations in John's proper frame. The equations and numbers are all worked out in Appendix 1, but this leaping time, the loss of simultaneity, is the key.

But isn't all motion relative? Who's to say who turned around? Can't we contend that it was John, not Mary, who turned around?

No, we can't. There is no disagreement about who turned around. It was Mary who fired her retrorockets; it was Mary who felt the acceleration. Both John and Mary know that Mary's proper frame accelerated and John's did not. In relativity, it is not true that "all motion is relative." What is true is that you can do all your calculations in any frame that moves at constant velocity. If the frame accelerates, then you have to take into account the leaping time of distant events.

The Tachyon Murder

The strange relativity result that the order of events can flip for different reference frames leads us into a new aspect of reality: the deep issues of causality and free will. These issues can be dramatized by the story of the tachyon murder.

A tachyon is a hypothetical particle that travels faster than the speed of light. Remarkably, relativity does not prohibit particles from traveling that fast. It says only that massless particles must travel *at* lightspeed, and that particles with a nonzero rest mass *cannot* travel at that speed (since the gamma factor would be infinite, and they would have infinite energy). The equations don't prohibit faster-than-light travel per se.

How can you get something moving beyond the speed of light without having it pass through that speed? The answer is that you can be *born* at super-lightspeed. Why not? Photons are not accelerated to lightspeed; they move at lightspeed from the instant they are created. So maybe we could create a tachyon that is already moving faster than lightspeed from its earliest existence. Such a scenario would not violate relativity. This is, indeed, the assumption made by physicists who search for tachyons.

Discover a tachyon, show it exists, and you will make physics history. Yet despite the upside of such a discovery, I decided many years ago not to bother searching for a tachyon. My reason borders on the religious. I believe that I have free will, and the existence of tachyons would violate that belief. Let me explain.

Imagine that Mary is standing 40 feet away from John. She has a tachyon gun that fires tachyon bullets that move at $4c$, four times the speed of light. She fires. Light moves at a speed of 1 foot per nanosecond (billionth of a second), so her tachyons move at 4 feet per nanosecond. In just 10 nanoseconds, the tachyon bullet enters John's heart and kills him. Let's assume he dies instantly.

Mary is brought to trial. She doesn't deny any of the facts I just described, but she insists on an unusual change of venue. She says she has a right to argue the case in whatever reference frame she chooses. They are all valid, the judge knows, so he allows her to proceed. She chooses a frame moving at half lightspeed, $\frac{1}{2}c$. Since that frame is moving slower than the speed of light, according to relativity it is a valid reference frame.

In the Earth frame, the two events (fire gun, hit heart) are separated by +10 nanoseconds. As I show in Appendix 1, the same two events described in a reference frame moving at $\frac{1}{2}c$ have a time separation of −15.5 nanoseconds. The negative sign means that the two events occur in the opposite order. The bullet enters the victim's heart before Mary fires the gun! Mary has the perfect alibi. John was already dead when she pulled the trigger. You can't murder a dead person. She expects to beat the rap.

The tachyon murder example is based on the same relativity principle that caused confusion in the twin and pole-in-the-barn paradoxes. If two events are sufficiently separated in space, and not too different in time, then there will be frames in which the order of events will reverse. Such distant events are called "space-like." Two events that occur near each other but separated in time are called "time-like." The order of space-like events depends on the frame of reference; the order of time-like events does not.

Once again, I refer you to Appendix 1 for the explicit calculation.

Is the tachyon murder scenario possible? How could analysis in the $\frac{1}{2}c$ frame be valid, if it has such an absurd implication? Does this mean that tachyons don't exist, or does it mean that relativity is nonsense? What if tachyons are really found?

Free Will Is Testable

One possible resolution for the tachyon murder paradox is that, in this world that has tachyon guns, Mary does not have free will. Even though she pulled the trigger *after* John died, she had no choice but to do so, since without free will, choice is illusory. All of her actions arise from influences and forces outside of herself. John died because it was inevitable that Mary would pull the trigger; the inevitability of physics created the combined scenario of shooting and death, and the order in which they occurred is irrelevant. There is no paradox if the world is governed by causal physics equations. The scenario presents a problem only if you think people have free will, if you believe that Mary could have decided not to fire the gun. If physics rules, then she does only what the various forces and influences on her cause her to do.

That's why I haven't searched for tachyons. I think that I have free will. Nothing in physics denies this, as long as tachyons don't exist (and the equations of relativity are valid). Of course, it is possible that my own free will is an illusion, and that I am really just a set of complicated molecules responding to local pushes and shoves. In that case, if I found a tachyon, I would go down in physics history, but I would be pretty unhappy to recognize that I couldn't take any credit for my discovery. It wasn't my doing.

On the other hand, it is intriguing that the concept of free will is scientifically falsifiable, at least in this manner. I'll talk more about what falsification means when we discuss the arrow of time, but for now, let me say that scientists generally agree that for a theory to be called scientific, you have to describe how you could prove it to be untrue. Some "theories," such as *intelligent design*, don't meet that standard. Remarkably, the theory that we have free will does. It has at least one falsifiable prediction: tachyons don't exist.

The paradox doesn't work for a sub-lightspeed bullet. I show in Appendix 1 that, if two events are separated by distance D in space and time T, then if D/T is less than lightspeed (that is, the bullet moves at less than the speed of light; the two events are *time-like*), the order of events is the same in all frames of reference. If you shoot someone with a real bullet, then changing reference frames will not help your case. In all frames, you fired before the victim died.

From time to time, one or another research team thinks it has seen a tachyon and issues a press release. That's what happened in 2011, when CERN, a large international research center near Geneva, announced that physicists there had observed some (not all) of the particles called neutrinos traveling faster than light. Headlines pronounced, "Tachyon Neutrinos Could be the Discovery of the Century." I didn't get excited. Such experiments are difficult and prone to subtle systematic errors. Indeed, less than a year later, CERN issued a second announcement retracting the claim and attributing the mistake to faulty electronics.

Tachyons, if they do exist (and we don't have free will), have some interesting properties. The gamma factor γ is imaginary (it's the square root of a negative number). We know energy is real (from Noether's theorem; see Chapter 3), so for real energy $E = \gamma mc^2$, the mass must be imaginary too. Tachyons have imaginary mass. That's OK; in Chapter 6 I'll describe how imaginary numbers aren't really imaginary; they are really real. But more curiously, as the speed of the tachyon gets higher and higher than lightspeed, as the tachyon velocity approaches infinity, the energy decreases! Zero-energy tachyons move at infinite speed. The energy of a tachyon approaches infinity as its velocity approaches the speed of light, from above, just backward from the way ordinary particles behave.

Incidentally, in the tachyon murder case Mary was convicted. When the judge declared her guilty, he explained that he had no choice; he had no free will, so he could do only what the forces on him made him do.

THE HEART OF EACH of these paradoxes is in the non-intuitive aspect of simultaneity. It is much easier to accept the idea that time slows down, or that moving objects get shorter, than it is to feel comfortable with the concept that when we say *now*, the word has no universal meaning.

Let's now address another paradox. It is a way to bring distant objects very close together in very short time. If you take distance and divide by the time you experienced, the resulting velocity—the rate at which the distant object came closer—can be much greater than lightspeed. Yet this behavior does not violate the theory of relativity.

5

Lightspeed Limit, Lightspeed Loophole

The distance between objects can indeed change faster than the speed of light . . .

It's the ship that made the Kessel Run in less than twelve parsecs!
— *Han Solo, in* Star Wars

Even though no ordinary object (one that you can bring to rest) can move faster than lightspeed, you can change the distance between you and a distant object at an arbitrarily fast rate, far exceeding the speed of light, without violating relativity. The paradoxical difference between *velocity* and *rate of change of distance* will prove important when I discuss the expansion of the universe and how that relates to the flow of time. I begin with the intimate connection between acceleration and gravity.

Einstein's Equivalence Principle

Some people are bothered by science fiction movies in which astronauts walk around their spaceships as if gravity were present. Some movies, such as *2001: A Space Odyssey* and *Interstellar*, have rotating wheel-like sections to give the astronauts simulated gravity. (Both of those movies correctly depict the rate of rotation needed to get an Earth-like gravity equivalent.) But the starship *Enterprise*, in *Star Trek*, has gravity without rotation. That bothers some people, but not me. Captain Kirk seems to have lots of energy available with his antimatter fuel, so I assume that, when in deep space, he keeps his ship accelerating at 1g—that is, at the same acceleration felt on the surface of the Earth. That gives him an artificial gravity that is completely Earth-like. The acceleration could be along the line of travel or perpendicular to it, depending on which starship surface he wants to stand on, which window he wishes to gaze out from.

Here's a curiosity about 1g acceleration. Do it for a year and, if classical physics were correct, at the end of that year you would exceed the speed of light. So 1g acceleration can get you going very fast. It makes a lot of sense for science fiction travel.

In fact, a year of 1g won't get you to lightspeed, because of relativity effects. We assumed constant 1g acceleration in the Earth frame. To produce a comfortable Earth-like equivalent gravity, we need to arrange for 1g acceleration in the reference frame that matches the proper frame of the rocket. If we use the relativistic formulas, it turns out that for acceleration a in our proper frame, the acceleration with respect to the Earth's frame will be given by a divided by the cube of gamma; that is, the acceleration is a/γ^3.

That formula is simple enough that with no further math, you can work out space travel conditions using a spreadsheet. Set up columns for time, location, and proper acceleration of 1g (a = 32 feet per second every second = 22 mph every second); columns for gamma, proper time interval (time interval divided by gamma), and acceleration in the Earth frame (a divided by the cube of gamma); and so on. Divide the time into short intervals and add up the small amounts of proper time to get the total proper time. You'll find some interesting results. In one year (of proper time) in a ship accelerating at 1g, you will have lightspeed of 0.76; after two years, 0.97; after three years, 0.995. Of course, you'll never quite get to the speed of light.

Suppose Captain Kirk decides to take a trip to Sirius. He doesn't use any special super-lightspeed drive but just keeps a comfortable proper acceleration of 1g. It will take him 9.6 years to get there, but he will experience only 2.9 years of aging. (I worked out these and the following numbers using a spreadsheet.) When he arrives, in his frame Sirius will be approaching at 99.5 percent lightspeed. Earth will be behind him but, because of space contraction, instead of being 8.6 light-years away it will be only 0.9 light-year distant. That's consistent with Kirk's experience of having traveled for only 2.9 years. If he wanted to stop at Sirius, it might have made more sense to accelerate at 1g for the first half of the trip and then decelerate at 1g for the second half.

Kirk experienced 2.9 years of time, yet the distance to Sirius changed by 7.7 light-years. That is a rate of distance change of 7.7/2.9 = 2.6 light-years per year, or 2.6 times the speed of light. This is what I call a *lightspeed loophole*. Distances measured in accelerating frames can change at arbitrary velocities. But the reason is that whenever you accelerate your proper frame, the distance to a faraway object can change with arbitrary rapidity. Jump your proper frame from one velocity to another, and the distance will suddenly be less, by a factor of gamma.

Achieving Lightspeed

Can you actually reach the speed of light? What would happen to time if you did? The lightspeed v/c would become 1. The time dilation/length contraction factor gamma would become infinite, seeming to suggest that when you reached lightspeed, your time would stop and your size (in the Earth frame) would shrink to zero. Moreover, since gamma would be infinite, your energy, γmc^2, would also be infinite. So yes, you could get to lightspeed if you applied infinite energy to yourself and accelerated for infinite time. Infinity is a lot bigger than all the energy in the universe, so that's not a practical solution.

Now let's look at some really high accelerations that have actually been achieved. BELLA is an electron accelerator built at the Lawrence

Berkeley Laboratory (where I've done most of my research). BELLA uses a laser to accelerate electrons, and its name is an acronym for "Berkeley Lab Laser Accelerator." It is only 3.5 inches long (9 centimeters), yet it can accelerate an electron to give it 4.25 GeV of energy in a few billionths of a second. GeV stands for a billion electron-volts; for comparison, the rest mass of an electron has energy mc^2 of 0.000511 GeV.

The length contraction factor for the electrons coming from BELLA is easy to calculate: it is just the final energy of the electron divided by its rest energy, since gamma = $\gamma = E/mc^2$. So, gamma is 4.25 GeV/0.000511 GeV = 8,317. BELLA is a wonderful accomplishment; it achieves enormous acceleration in a compact space. It was a very long and difficult path developing this "simple" apparatus.

Point BELLA at the star Sirius, 8.6 light-years away. In the proper frame of an electron just injected into BELLA, that is indeed the distance to Sirius. A few billionths of a second later, the electron is moving with gamma = 8,317. That's 0.9999999927 times lightspeed. In its proper

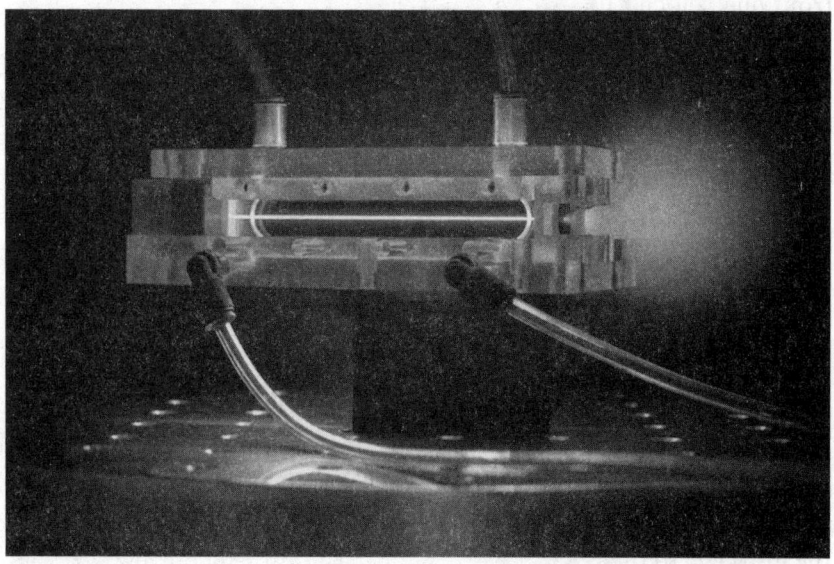

Figure 5.1. BELLA, the device built at the Lawrence Berkeley Lab that can accelerate an electron to 99.99999927 percent the speed of light in 3.5 inches.

frame, Sirius is 8,317 times closer, only 0.001 light-year away. The distance between Sirius and the electron, measured in the electron's proper frame, decreased by nearly 8.6 light-years in about a billionth of a second. That rate of change in distance is more than 8.6 billion times the speed of light.

This example shows that distances *measured in accelerating frames* can change with arbitrarily high velocity, 8 billion times lightspeed or more. Such rapid change of distance will prove important in general relativity, since it treats gravity as acceleration. This super-lightspeed phenomenon will lead to some very important effects in cosmology. In particular, in the standard formulation of the Big Bang theory, the galaxies are not moving but the distance between them is increasing. That rate of change in distance is not limited by lightspeed—a behavior that will prove important when we discuss the theory of *inflation*, which involves a very rapid expansion in the size of the universe. And we will postulate (in Part V) that the expansion of space is accompanied by an expansion of time, and that such expansion will account for both the flow of time and the meaning of *now*.

Time Flows Faster Upstairs

Gravity also affects time. Live upstairs and your life will move ahead faster than if you lived on the lower floor. This phenomenon is not in dispute. Just like the time dilation of velocity, the time speedup of altitude affects our GPS satellites (it is bigger than the velocity effect) and must be taken into account to give us accurate locations.

The link between time and gravity was another amazing prediction of Albert Einstein. It came from his physical intuition that gravity should be indistinguishable from an accelerating reference frame—an assumption that he called the *equivalence principle*.

Captain Kirk experienced the equivalence principle in his artificial gravity. Acceleration is a great simulator for gravity. You feel the equivalence principle in an old elevator that starts downward (too) quickly; for an instant you feel as if you weigh less. You experience the equivalence principle in the Star Tours ride at Disneyland. As you sit in the closed room, your views of the "space station" are visible through the windows; the windows, of course, are just video screens. Then the ride suddenly accelerates; you are pressed back against your seat as the scene outside rushes backward.

It is a completely compelling illusion. You feel yourself being accelerated, just as you do when a plane accelerates down a runway or the driver of a car steps hard on the gas pedal. But, of course, you are not accelerating. As the video shows the outside whizzing past, the hydraulic lifts that hold the room have just tilted it backward by about 30 degrees. Gravity is what is pulling you against your backrest. But because you see the world flying by in the video windows, the illusion is persuasive. Disneyland uses the equivalence principle of Einstein. Gravity and acceleration are indistinguishable.

Because gravity is *just* acceleration, Einstein could use his equations for accelerating reference frames to calculate the effects of gravity. He did this and more, building general equations that could accommodate even complicated gravitational arrangements, such as those of stars and black holes. But his work was fundamentally that of the equivalence principle: gravity is indistinguishable from acceleration.

One result of his theory is the one I mentioned, that time moves faster upstairs. Again, Einstein's equation is remarkably simple. I derive it in Appendix 1. The speedup factor is given by $1 + gh/c^2$. The number 1 represents the normal rate of time flow; it's the second term, gh/c^2, that makes time go faster. Here h is the height, g is the acceleration of gravity (32 feet per second every second), and c is the speed of light.

Let's plug in some numbers. I'll use units of feet and seconds. Make h the height of one flight of stairs, about 10 feet; g is 32, so gh is 320. The velocity of light is 1 foot per nanosecond, a billion feet per second; that makes c^2 a billion billion. So, gh/c^2 is 320×10^{-18}. A day has 86,400 seconds, so that amounts to 0.27 nanoseconds per day.

In 1915, when Einstein published his original papers on this gravitational time effect, it was too small to detect. It remained elusive for decades. Then, in 1959, to the amazement of the entire world, this tiny change was actually observed and measured by Robert Pound and his student Glen Rebka. They were able to send a gamma ray downward 74 feet and detect the change in frequency by using a recently discovered phenomenon known as the Mossbauer effect.

The term gh/c^2 assumes constant gravity. The equations become only a little more complicated when the strength of gravity changes with altitude, as happens when you get far above the surface of the Earth. But for the special case when you would like to know how much slower time runs on the surface of Earth or on some other planet than it does in distant space, instead of using gh/c^2, just use gR/c^2, where R is the radius of the planet and g is its surface gravity.

As I mentioned already, this time effect gets pretty large for GPS satellites. They orbit at an altitude of about 12,500 miles, so far above the surface that we are essentially comparing clocks on the surface of the Earth to those in far space. So the appropriate equation is gR/c^2. Plug in the numbers and you'll find that the clock on Earth is running slower by 0.7 parts per billion than the one in space. That adds up to 60 microseconds per day, which would give a distance error of 60,000 feet, about 11 miles.* And this error would double to 22 miles by the second day, if the clocks didn't take the gravity effect into account.

You can look up the radius and the surface gravity of various planets and stars and calculate gR/c^2. Compared to a clock in space, time on the surface of the sun runs slow by 6 parts per million, and for a white dwarf star, by 1 part per thousand. Time comes to a complete halt at the surface (Schwarzschild radius) of a black hole. This last result is fascinating and will prove important as I discuss the nature of black holes later in the book.

---

* If you take into account the fact that the satellites aren't quite at infinity, you get a slightly smaller mileage error per day—not 11 miles, but 8.6 miles.

The movie *Interstellar* depicts the time dilation near a black hole in an interesting way. A group of astronauts travels down toward a black hole—not all the way in, but pretty deep. (You can, in principle, return as long as you don't reach the Schwarzschild radius.) Meanwhile one astronaut remains orbiting above. When the traveling astronauts return, after just a few days, the astronaut in orbit has experienced twenty-two years. They are trying to save the Earth, but they know that their time is moving very slowly compared to that of the outside world, and that the growing ecological disaster on Earth (in that movie) is proceeding at a much more rapid pace than the time they had experienced. Time dilation is their enemy and gives their work an incredible urgency. It also means that when (and if) they return, their children will be older than they are. (I don't particularly recommend the plot in *Interstellar*, but the time dilation effects are accurate, vivid, and memorable.)

EINSTEIN SHOWED THAT the time of an event affects its location, and the location of an event affects its time. But it was his former math teacher who first recognized that what Einstein had done, without realizing it, was to *unify* space and time, to deduce that time and space were no longer separate, but each was part of space-time.

6

Imaginary Time

The concepts of time and space are unified . . .

There is a fifth dimension, beyond those known to man.
— *Rod Serling,* The Twilight Zone

After Einstein published his initial relativity papers, his former math teacher Hermann Minkowski expressed astonishment. He hadn't remembered Einstein as a particularly outstanding student. (It is wrong to say, as some suggest, that Einstein was "bad" at math; Minkowski's class was advanced.) But Einstein's relativity papers were revolutionary, amazing, and rock solid. Those papers changed Minkowski's life.

Minkowski then made an astonishing leap himself—one that had enormous influence back on Einstein, one that led ineluctably to the equations on which the modern theory of the universe is based, Einstein's general theory of relativity.

Einstein's original relativity equations had linked space and time. According to his math, the time of an event depended not only on the time in another frame, but also on its position. Minkowski took Einstein's equations and did something that might have looked to some like a mathematical trick, but it had a profound implication. He formulated relativity theory in a clever way in which space and time were coordinates in a four-dimensional "space-time." But to accomplish this, he had to make the time coordinate imaginary.

Imaginary time? By that, I mean that an event is designated by four numbers: x, y, z, and it, where i is $\sqrt{-1}$ and t is time. Why do such a crazy thing? Minkowski's reason was that it turned this combination of coordinates into a mathematical object we call a *vector*, one with extremely useful properties.

Some might think that making time imaginary for mathematical advantage is throwing out the baby with the bathwater. We know that time is real. To treat it as imaginary seems crazy. But to physicists and mathematicians, *imaginary* numbers are not imaginary at all, at least not in the sense that the tooth fairy is imaginary.

It is worth confronting the issue of *imaginary*, since it shows up not only in relativity, but also in quantum physics, in which the quantum wave is a combination of real values and imaginary ones. For states that have well-defined energy, in quantum physics time is again combined with $\sqrt{-1}$ in an exponential that gives the time dependence. So let's talk about imaginary numbers.

Zero, Irrational, and Imaginary Numbers

To understand imaginary time, it helps to realize that the term *imaginary* as used in physics and math does not have the same meaning as when it is used in literature and psychology. Ironically, use of the word *imaginary* in mathematics simply reflects mathematicians' lack of imagination. Like physicists, mathematicians tend to use ordinary words to mean extraordinary things. They don't have sufficient imagination to come up with a new word, so they steal a common word and ascribe a new and specific meaning to it. Many scientists do that.

Excuse me for a moment while I rant about *sciencespeak*. I ask, by what right does a scientist tell us that the American buffalo is not a buffalo? Or that a spider is not an insect? Or that Pluto is not a planet? Scientists attempt to hijack these words and then tell us when we can and cannot use them. They didn't make up these words, so they don't have the right to narrow their definitions. In my mind, an American buffalo is an American *buffalo*. In the 1600s, not only spiders but even earth-worms and snails were called insects. I once had a mathematician tell me that I couldn't tie a knot in my shoelace, because by mathematical definition, anything that can be untied is not a knot!

No one gave scientists and mathematicians the right to change the meanings of common words. A wonderful conclusion, based on this logic, is that Pluto is still a *planet*! I took a vote in my class, and it came out 451 to 0 to keep Pluto as a planet. Since that poll had more partic-ipants than did the one taken by the IAU (International Astronomical Union), I conclude that our vote should carry the day. Nobody gave the IAU authority to decide. (And I am a member of the IAU.) Pluto is still a planet. End of rant. Back to imaginary numbers.

In my education, I've seen very smart students reach their tolerance limit in math when presented with imaginary numbers. How can they work with something that doesn't exist? When they encounter imagi-nary numbers, they seem to feel that math has become too abstract, too far removed from reality for them to be able to understand it.

In the spirit of opposing sciencespeak, I announce that imaginary numbers are not imaginary. In fact, $\sqrt{-1}$ does exist. To understand how, let's look at some other abstract numbers. Does the number 0 "exist"? The Romans said no. They found it self-evident that nothing does not exist. As a result, Roman numerals have no way of writing 0. A Roman subtracting IV from IV would just leave a blank space. But how can a blank space be distinguished from a failure to solve the problem? The concept of using a symbol for *nothing* was a leap of imagination that Romans never made (unless you consider Ptolemy to be Roman). I sus-pect some mathematicians (or maybe accountants) at the time argued for such a symbol, just because it would be useful, but it was concep-tually difficult to put a symbol on something that was nothing. Zero doesn't exist, does it? It's just in your imagination, right? It's imagi-nary, right?

The Greeks were amazingly mathematically sophisticated. Archi-medes showed that the volume of a sphere is $4/3 \, \pi R^3$. Try deriving that yourself without calculus! Still, they did not have a symbol for zero, at least not until AD 130, when Ptolemy in Alexandria made limited use of one. They, too, just left the space blank.

I used to have fun with my daughter when she was five years old. I would ask her, Who is sitting in the backseat? She would say, "Nobody." Is nobody's window open? "No." But my window is open! So how could you say that nobody's window is open? "Daddy!!!" As annoyed as she was, she would immediately turn the wordplay back on me. She loved this game, never realizing that I was preparing her for abstract math.

What about negative numbers? I recall a seventh-grade math teacher (she was the worst teacher I ever had) who told our class that negative numbers don't exist. "Just pretend they do," she said. Fortunately, I was precocious and decided she was wrong. I remember thinking to myself, *a negative number is like owing something.* But I think her recommended approach was the end of math for half the kids in the class. They were never going to feel comfortable manipulating things that didn't exist. Negative numbers did exist, for me.

So in seventh grade I had already figured out that numbers aren't things, but rather concepts useful for calculating. Does any number exist? Or are numbers just abstractions that we use to organize our thoughts? This is actually a philosophical question on the meaning of *existence* that has filled essays and books. (I have a book on my desk now titled *Does Santa Exist?* It is a serious book about the meaning of the word *exist*.) We'll return to this issue when we discuss some recent concepts in physics that may or may not "exist." One of these will be the quantum wave function. Another will be the Schwarzschild surface of black holes.

The ancient Greeks believed (that's the right word) that for numbers, only integers existed. They held this truth to be self-evident. They thought that all other numbers could be written as fractions, ratios of integers, such as 22/7. Pythagoras was given credit for discovering that the tones of music are such ratios; an "octave" meant an exact factor of 2 (in the length of a vibrating string). It is called an octave because it spans eight notes. A fifth in music, spanning five notes, came from a string length factor of 3/2. A fourth was a factor of 4/3.

Then a startling thing happened, not only for the history of mathematics, but for the human understanding of reality. The Pythagoreans, about 600 BC discovered that $\sqrt{2}$ could not be written as a ratio of integers. As a result, they called $\sqrt{2}$ *irrational*. Not rational. Crazy.

This may sound like an arcane mathematical matter, but think about it. How could you ever be sure that statement is true? After all, $\sqrt{2}$ is not a particularly weird number; it is the length of the hypotenuse of a right triangle whose arms each have length 1. Physical measurement could not possibly conclude that the number is irrational. You could never try all possible integer combinations. Suppose I told you that $\sqrt{2}$ = 1,607,521 divided by 1,136,689. It isn't, but that fraction is very close. Try it; do the division on your calculator and then square it. Or use a spreadsheet.

In discovering the irrational nature of $\sqrt{2}$, the Pythagoreans took a major step in recognizing the reality of nonphysics knowledge. I give a proof of the irrationality of $\sqrt{2}$ in Appendix 3. It's not hard; take a look at it. I'll talk more about $\sqrt{2}$ later in this book, but let's continue with our investigation of the meaning of *imaginary*.

At least $\sqrt{2}$ could be constructed using a straight edge and compass. As I've already said, it is the length of the hypotenuse of a right triangle with unit-length arms. But the ratio of the circumference of a circle to its diameter, the number we call π, could not be so constructed. It turns out to be even stranger than $\sqrt{2}$; we call it *transcendental*, as in transcendental meditation.

One even more astonishing fact about the irrationality of $\sqrt{2}$— something that shows how extraordinary this fact truly is—is that it was discovered only once in the history of civilization. All other statements of this fact around the world can trace the origin of their knowledge to the work of the Greek mathematicians.

What about $\sqrt{-1}$? It is not an integer, it is not a rational number, it is not an irrational number, and it is not transcendental. Does that mean it doesn't exist? In a way, it doesn't exist, but only to the extent that no numbers really exist. They are tools we use in our minds to do computations. If a tool (such as 0 or –7 or $\sqrt{2}$) is useful, then use it. The fact that $\sqrt{-1}$ is not on the previous list of weird noninteger numbers does not mean it doesn't exist. In my mind, in the minds of mathematicians and physicists, it is as real as the number 1.

The main problem with imaginary numbers is their name. If $\sqrt{-1}$ had been called "extended" instead of imaginary, perhaps it would not have given as many qualms to generations of students. Or maybe we should call it the "E-number," after the great mathematician Leonhard Euler, who showed us that $e^{\pi\sqrt{-1}} + 1 = 0$. Richard Feynman called this equation "the most remarkable formula in mathematics." It relates five important numbers—e (the base of the natural logarithm), π, $\sqrt{-1}$, 1, and 0—in a most unexpected way, and in a way that turns out to be enormously valuable in both electrical engineering and quantum physics. Alas, Euler already has e, the base of natural logarithms, named after him.

Now back to imaginary time. Clocks show no $\sqrt{-1}$; they have only a set of integers, plus a big hand and a little hand. How can time be imaginary, or even extended?

The answer is that in Minkowski's formulation, time is still given in hours, minutes, and seconds—real numbers. The thing that is imaginary is the abstract space-time created by Minkowski. Time remains real, but the coordinate that became part of space-time is the real number t multiplied by the imaginary number $\sqrt{-1}$. Nonetheless, when talking about Minkowski's construct, 4D space-time, physicists refer to the combination it as "imaginary time."

Imaginary Time and 4D Space-Time

The most enduring of Minkowski's contributions was not formulating imaginary time, but introducing the concept of *space-time*. He showed that the equations used in relativity to calculate the position and time coordinates in a new reference frame could be thought of as a rotation in space-time. Theoretical physicists found that idea very appealing. Rather than having only an equation, they could think of relativity in terms of pictures. True, they had to think in 4D pictures, and some of them could do that, but most physicists tried to reduce their problems so that they had only one space dimension (like the straight-line route that Mary took between the Earth and the star) and one time dimension. Then the space-time diagram could be drawn on a piece of paper, and a change in coordinate systems, from one reference frame to another, meant simply rotating the picture.

The initial significance of space-time was that it turned relativity from an algebraic problem into a geometric one. The influence on Einstein was dramatic. He explored the possibility that all equations of physics were just complicated geometry. He started with gravity, since he had already concluded that a gravity field was equivalent to a uniform acceleration. From this he deduced that time flows faster upstairs than downstairs. Could all of gravity, not just uniform gravity fields, be turned into geometry? What about electromagnetism?

The Most Splendid Work of My Life

Einstein worked for a decade to develop a geometric understanding of gravity. It was one of the most fantastic episodes in the history of human thought. When he completed his investigation, he had taken space-time and allowed it to have arbitrary geometry, including curves and stretching. Just as the surface of the Earth has mountains and valleys, the four dimensions of space-time could twist and turn, compress and expand, but still remain continuous and smooth. Planets and satellites that appeared to orbit massive bodies were, in this view, simply following their noses, moving in what they felt to be "straight lines" (called *geodesics*). The old gravitational field of Newton was gone, replaced by variable geometry that depended on the density of nearby energy (including mass energy).

Einstein succeeded in finding an equation in which the geometry of space-time was determined by its energy content. In this approach, gravitational force does not exist. The presence of mass means the presence of energy; the presence of energy distorts space and time; the distortion of space and time means that objects will appear to be responding to gravitational forces, when in fact they are only following their noses (that is, moving straight ahead) through a complicated curved space-time. In this language, planets orbiting a star are actually moving in a straight line—a straight line not through space, but through space-time.

By 1915, Einstein had not only found his final equation, but also had convinced himself (and soon the world) that it was right. The equation looks simple:

$$G = kT$$

where $k = 2.08 \times 10^{-43}$ in standard physics units (meters, kilograms, and seconds).

That is *the* equation of general relativity! All the complexity is hidden in the definition of the two terms G and T. We now call the quantity G the *Einstein tensor*; it is a mathematical structure that describes the local curvature and density of space-time. What does that mean? Space is no longer simple. Because space can be expanded and contracted, you can, for example, squeeze a lot of space into a small region. The same is true for time; that's how the equations handle time dilation. If a nearby region contains a black hole, then you might discover that just to cross from one side to the other you would have to travel an infinite distance. It's like crossing a mountain; the straight-line distance involves not just forward motion but a lot of up and down. But in Einstein's theory, there is no mountain-like up and down; rather, there is just more space and more distance squeezed into the region.

In the equation, the quantity T describes the energy and momentum density of space.* The equation says simply that the local geometry of space and time is determined by the local energy content, described by T.

---

* In this formulation, not only space, but also energy, has four components, consisting of the energy and the three components of momentum. T is called the "energy momentum tensor" but for simple weak fields, it is just the energy/mass density.

In fact, within a constant, G and T are equal. Empty space is described by $G = 0$, although this equation does not mean that empty space always has a simple geometry; it means only that the curvature of space has relatively simple behavior. Einstein's equation tells us not only about the gravity of the Earth and the sun, but about the gravity of black holes and the universe. Buried in the solutions of this equation is the possibility that the entire universe is finite in extent, or infinite, that space can expand and contract, that black holes have time within that corresponds to beyond infinity on the outside (see the next chapter).

Perhaps most remarkably, by stretching and squeezing both space and time, Einstein could have objects accelerating even though their locations were not changing. You, sitting at the surface of the Earth, are (in this geometry) constantly accelerating upward but not moving. That upward acceleration is what we call the gravitational pull of the Earth, and that acceleration can be thought of as responsible for the gravitational time effect.

Many people mistakenly think that squeezing extra space between objects requires a fifth dimension, beyond the four that we know. They mistakenly think of the extra space as resulting from a mountain-like structure, bending off into the fifth dimension, lengthening your otherwise simple path. Such a fifth dimension could exist, but the math doesn't require it. Space is not a solid; the amount of space in a region is not fixed. You don't have to imagine that there is an external dimension to describe the complicated "geometry" of relativity. You simply need to recognize that distances and time intervals are flexible, just as they were in 1905 relativity. Even back then, a 40-foot pole could be squeezed into a 20-foot barn by using the space contraction (at least in theory), with no need for a hidden dimension to fold up the pole.

Notably missing from the equation of general relativity is $\sqrt{-1}$. In the end, Einstein found (and developed himself) a mathematical way of approaching space-time that didn't involve imaginary numbers. He didn't exclude $\sqrt{-1}$ because it is unphysical (it isn't); he found a different approach, using something called non-Euclidean Riemannian geome-

Figure 6.1. Albert Einstein, in 1921.

try, that made the calculations more elegant, more powerful, easier to apply to new situations, and easier to interpret.

And for weak gravitational fields, such as the ones around the sun (black holes have strong fields), Einstein's equations were usually indistinguishable from the old gravity equations of Newton, who said that the acceleration of gravity from a mass M was given by $a = GM/r^2$. Newton's equation was just an approximation (although an extremely good one) for the truer Einstein equation of general relativity. Niels Bohr, one of the founders (along with Einstein) of quantum physics, later called this property the *correspondence principle*. New theories must produce the same results as the old theories, in the domains in which the old theories were successful. For general relativity, that meant slow velocity and relatively weak gravity.

But there were differences between the new theory of gravity and Newton's old theory. With his new equations, in 1915 Einstein calculated that the orbit of a planet such as Mercury around the sun would not be a simple ellipse, but an ellipse that gradually shifted its axis. Einstein's calculations explained a known conundrum that had puzzled scientists for fifty years. The orbit of Mercury had been observed to have a shift, called the advance of the perihelion. With no adjustment, no added numbers, Einstein's equations accurately accounted for the advance. It was a post-diction, not a pre-diction, because the advance of the perihelion had been known since 1859.

I find it hard to imagine how Einstein felt when he first calculated the orbit of Mercury and found that his results matched the well-known but unexplained shift. He found this result in 1913, working with a collaborator, Michael Besso. Einstein wrote in a letter to his friend Hans Albert, "I have just completed the most splendid work of my life." That's quite a statement for the man who had already invented ordinary relativity, proved that atoms exist by his interpretation of Brownian motion, and created the foundation of quantum physics with his paper on the photoelectric effect.

In his 1915 work, collected together in an extraordinary and historic 1916 paper, Einstein made two other predictions. He said that starlight passing close to the sun would be deflected by about 1.75 seconds of arc. Within a few years, Arthur Eddington, a physicist I will talk about a lot

Figure 6.2. Calvin describes the curvature of space.

in this book, had verified that prediction by making difficult measurements during a total solar eclipse. Eddington's verification catapulted Einstein to international fame. Einstein's prediction that time would flow more rapidly at high altitude took longer, but it was verified forty-four years later by Pound and Rebka.

Space-Time

Once Minkowski and Einstein had introduced the concept of space-time, all sorts of other physics things became readily interpreted as having four dimensions. Energy and momentum, previously seen as related but separate concepts, became the components of a 4D object: the three components of momentum, in the x, y, and z directions, became three components of the 4D energy-momentum vector, and the fourth component was the energy. Einstein "unified" momentum and energy in the same sense that he (and Minkowski) had unified space and time.

Other physics quantities all fell into this "beautiful" mathematical formulation. Electric and magnetic fields were no longer separate quantities; they were just separate components of a 4D object known as a *tensor*. Amazingly, if you rotated the coordinates, you would turn electric fields into magnetic fields, and vice versa. The math of the rotation was essentially identical to that of the Lorentz/Einstein transformation; in the jargon that developed, this property was called relativistic "covariance." These rotations turned out to be mathematically equivalent to the classical "Maxwell equations" that related electric to magnetic fields, the equations that were used to design motors and electric generators. It was a wonderful step toward the unification of physics.

Einstein continued his amazing productivity. Soon after he finished his initial papers on general relativity, he wrote several papers on the emission of radiation in which he predicted a previously unknown phenomenon he called *stimulated emission*. His work led directly to the invention of the laser by Charles Townes in 1954. The world *laser* is an acronym for "light amplification by *stimulated emission* of radiation."

Einstein considered his original 1905 theory of relativity to be the first step in a program of understanding all of physics via geometry. Through the equivalence principle, he had included gravity. He did not think he was done. He wanted to turn electromagnetism theory into a geometric theory, similar to what he had done with gravity, and combine it with general relativity. In 1928 he began a series of papers on a "unified field theory" that attempted to do just that. Most scientists today regard Einstein as having finally taken the wrong path, probably because he did this work without including the quantum physics that he had helped uncover.

By incorporating quantum physics, many theoreticians believe they recently approached Einstein's goal of a unified theory, although it is not based on geometry. The approach, called *string theory*, combines general relativity and quantum physics, bringing together into one subject the forces of gravity, electricity, and magnetism; the "weak" interaction that causes radioactive decay; and the "strong" force that holds the protons together in the nucleus despite their enormous mutual electric repulsion.

String theory has generated enormous enthusiasm, and there are many popular books on the subject. My evaluation is that string theory is not the solution we are looking for. It has made many predictions (for the existence of new particles) that have not been verified, and it has made no predictions that have turned out to be correct. Some have claimed that the strongest evidence for string theory is that it is mathematically consistent, with no arbitrary (and difficult-to-justify) computational tricks to get rid of the infinities that arise in standard quantum physics. Some say its greatest achievement is that it "predicts the existence of gravity." Of course, gravity was known long before string theory, but the remark reflects the fact that string theory *requires* the existence of the relatively weak (compared to other forces) gravitational field.

EVEN WITHOUT FURTHER theoretical additions, shortly after Einstein published his work surprising phenomena were discovered within the general theory of relativity itself. The theory could be applied to the universe; it could be applied to very dense objects. Robert Oppenheimer, the future director of the Manhattan Project and the "father" of the atomic bomb, is the person usually given credit for using the equations of relativity to show that a black hole might actually be created when a heavy star collapses. Indeed, there appears to be a black hole that is only (in astronomers' usage) 6,000 light-years away from the Earth. Theoretical study of black holes has forced us to think of time in a new way, a way that challenges many of our innate prejudices.

7

To Infinity and Beyond

Time near black holes is far stranger than most people imagine . . .

To infinity and beyond!
— *Buzz Lightyear,*
in Toy Story

Physicists are often puzzled by their own equations. It's not always easy to spot the implications, even the most dramatic ones. To help understand their own math, they look at extreme cases to see what happens. And no extreme situation is more extreme in this universe than the extreme of black holes. Looking at black holes gives us important insights into some very peculiar aspects of time.

If you orbit a small black hole (mass of the sun) from a reasonable distance—say, a thousand miles—you'll feel nothing in particular. You are in orbit about a massive object that you can't see. Since you are in orbit, you feel weightless, just as all orbiting astronauts feel. You are not getting sucked in; black holes (despite popular science fiction) do not draw you in. If you were orbiting the sun, at this close distance you'd be inside it and you'd burn to a crisp in a millionth of a second, but the black hole is dark. (Microscopic black holes radiate, but very little comes from big ones.)

The distance around your orbit is 2π times your radial coordinate value of 1,000 miles. If a friend is orbiting the hole, but on the other side, going the opposite direction, then you'll meet after you each go a quarter orbit. But when your friend is diametrically opposite you, the straight-line distance between you is infinite. There is a lot of space near the black hole.

If you fire your retrorockets, stopping your orbital motion, you will indeed be pulled into the hole, just as you would be pulled into any massive object. (The way satellites de-orbit is precisely that: firing retrorockets and then letting gravity pull them in.) Before ten minutes pass in your proper frame, before you are ten minutes older, you will reach the surface of the black hole, at the *Schwarzschild radius* (discussed in Chapter 3). Now for some astonishing results concerning time. When you hit that surface, ten minutes after beginning your fall, the time measured on the frame of the orbiting station will reach infinity.*

That's right. It takes infinite time to fall into a black hole, measured from the frame of someone outside. From your accelerating frame falling in, it takes only ten minutes. At eleven minutes, the time outside has gone to infinity and beyond.

That's absurd! Possibly, but in classical relativity it is true. Of course, there is no way for you to experience the potential paradox, because beyond infinity is the time on the outside, and once you have entered the black hole you are there forever. There is no measurable contradiction. This is an example of what physicists call *censorship*. The absurdity is unobservable, so it isn't really an absurdity.

Are you satisfied with that "beyond infinity but censored" answer? I suspect not. I find it mind-numbing. But everything about time I find mind-numbing. We'll encounter another absurd but censored result with quantum wave functions and entanglement. These examples challenge our sense of reality, and leave an unsatisfied feeling. As Nietzsche said: When you gaze long into an abyss, the abyss also gazes back into you.

---

* L. Susskind and J. Lindesay discuss this infinite fall time in *An Introduction to Black Holes, Information, and the String Theory Revolution* (2005), p. 22. They station "Fidos" observers along the fall path who watch the object fall and report to the outsider. "According to this viewpoint, the particle never crosses the horizon but asymptotically approaches it." Quantum theory could conceivably change this conclusion.

Black Holes Don't Suck

Let's return to my statement that black holes don't suck you in, that you orbit a black hole just as you would orbit any other mass. Suppose Mercury were orbiting a black hole that had the same mass as the sun has. How would the orbit be different? According to popular belief, the black hole would suck the tiny planet in. According to general relativity, there would be no difference in the orbit. Of course, Mercury would no longer be hot, since the intense radiation of the sun would be replaced by the cool darkness of the black hole.

Mercury currently orbits the sun at a radial distance of 36 million miles. Suppose you orbited the sun at 1 million miles from its center, just above the solar surface. Aside from the heat, and possibly drag from the solar atmosphere, you would cruise around in a circular orbit and return to your starting point in about ten hours. Now replace the sun with a solar-mass black hole. You would still orbit in ten hours. The gravity, at that distance, would be identical to that of the sun. You have to get very close to a black hole before you notice any special effects. As with any star, the closer you get, the faster you have to move to stay in a circular orbit. As a rule of thumb, you won't see much difference until you are so close that your orbiting speed is approaching the speed of light.

For the sun, the maximum gravity is at the surface, just as it is for the Earth. Go below the surface, and the mass that attracts you, the mass below you, is less than at the surface. At the very center of the sun, the gravity is zero.

For a black hole, however, the surface is close to the center. From the Schwarzschild equation I gave earlier, the radius of a solar black hole can be calculated to be about 2 miles. At a distance of 10 miles, to stay in orbit you would have to move at one-third the speed of light; your orbital period would be one-thousandth of a second. In those conditions, we have to use relativity to do the calculations.

Reaching Lightspeed and Going beyond Infinity

When you get in close to a black hole, time progresses very slowly, and even though the distance around the orbit may be small, there is a lot of space between you and the hole. The space is conventionally depicted to physics students with a diagram such as the one in Figure 7.1. Think of this diagram as representing a black hole in 2D space (the surface). The black hole itself is at the center, below where the curved space is pointing.

This is a useful diagram, but it is somewhat misleading, because it implies that space has to curve into another dimension (for this diagram, that's the dimension that goes downward) to accommodate the enormous distances close to the black hole. In fact, no such dimension is needed; the space simply gets compressed from relativistic length shortening. The diagram is also used in popular movie depictions of black holes. When Jodie Foster falls into a wormhole in *Contact*, it looks very much like the diagram in Figure 7.1. (Wormholes look like two almost-black-holes connected before the Schwarzschild radius is reached; fall in

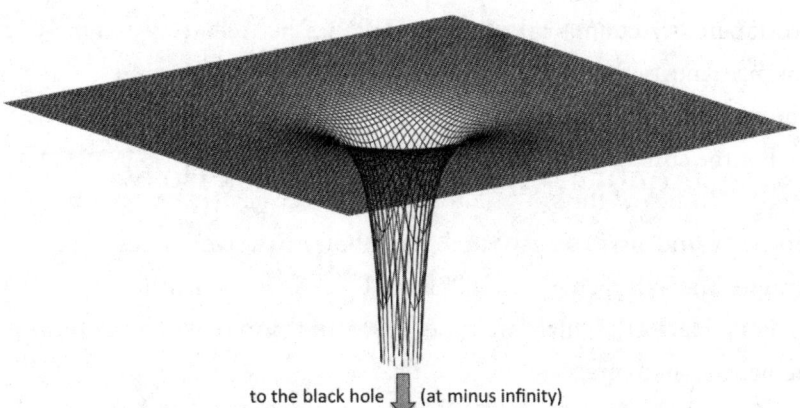

to the black hole ⬇ (at minus infinity)

Figure 7.1. Depiction of a 2D black hole. The distance to the black hole, measured by the time it takes light to reach it, is infinite, even though the distance to go around it is the same as in ordinary space.

one, fly out the other.) In fact, a black hole would not look at all like the diagram. Unless other things were falling in with you, it would just look like a totally black sphere.

With that caveat, the diagram is useful. It illustrates the basic features of black holes and can be used to answer some simple questions, such as: How far is it from the outside (the relatively flat region) to the surface of the black hole? The answer is infinity. Measure along the falling surface into the hole and you go down forever. You'll hit the radius of the black hole only at the bottom, but that is infinitely far down.

If it is infinity to the surface of the black hole, what did I mean when I said you were 10 miles distant? I confess I was being misleading. I was using the conventional coordinates. The radial coordinate r is defined by saying the distance around the black hole is $2\pi r$, just as in ordinary space. In Figure 7.1, the conventional x, y coordinates are represented by the grid lines. Note how far apart they get in the hole; the large distance between them shows that there is a lot of space in there. Physicists use these conventional coordinates in the equations, but they keep in mind that the distance between the 3-mile marker and the 4-mile marker might indeed be 1,000 miles. Because conventional geometry does not work here, we cannot calculate the distance between two points by just taking the difference in the coordinates.

Actually, There Are No Black Holes

You can find lists of suspected black holes in astrophysics books and online. The Wikipedia article "List of Black Holes" identifies more than seventy. Here's the catch: we have reason to think that none of these are actually black holes.

The way an astronomer identifies a black-hole candidate is to find an object that is very massive, typically several times that of the sun, yet is emitting little or no radiation. Some of the candidate objects emit bursts of X-rays, which are thought to indicate that a chunk of matter (a comet? a planet?) is falling in, and as it does it gets ripped apart and heated by the large differences in gravity across its own body, enough to emit X-rays. Other candidates, called supermassive black holes, contain hundreds of millions of solar masses.

One such supermassive object exists at the center of our own Milky Way Galaxy. We see stars orbiting very close to this center and moving and accelerating very rapidly, indicating the presence of a very large mass. But there is no light, so whatever is pulling these stars is not a star itself. Physics theories suggest that such a large accumulation with no emission can only be a black hole.

Why do I say that there are no true black holes on the list? Recall the calculation showing that it takes infinite time to fall into a black hole. A similar calculation shows that it takes infinite time to form a black hole, measured in our time coordinate. All that material has to fall, effectively, an infinite distance. So unless the black holes already existed at the moment the universe was created, unless they were primordial black holes, they haven't yet reached true black-hole status; there hasn't been enough time (from our outside proper frame) for the matter to fall the infinite distance that characterizes a true black hole. And there is no reason to think that any of the objects are primordial (although some people speculate that one or more might be).

I am being somewhat pedantic. It takes forever to fall into a black hole, but you get in pretty far in just a few minutes, in your own proper time measured by your own co-falling watch. From the outside frame you'll never reach the surface, but you'll be turned into a crepe-like object in relatively short order. So, in some sense, it hardly matters. That may be why, in 1990, Stephen Hawking decided to pay off his 1975 bet with Kip Thorne and concede that Cygnus X-1, the X-ray source in the constellation Cygnus, was indeed a black hole. Technically, Hawking was right, not Thorne. Cygnus X-1 is 99.999 percent of the way to being a black hole, but it will take (from the reference frame of Hawking and Thorne) forever to go the rest of the way.

One particular quantum loophole might circumvent my statement that black holes don't exist. Although it takes forever for a black hole to form in Einstein's original general theory of relativity, it doesn't take too long for one to "almost" form. The time from when the falling matter reaches twice the size of the Schwarzschild radius, to when it reaches within a tiny distance at which quantum effects are big (called a *Planck distance*, something we'll discuss later), is less than a thousandth of a second. At that point, we don't expect the ordinary general theory of relativity to hold up.

What happens next? The fact is, we don't really know. Many people are working on the theory, but nothing yet has been observed and verified. It is interesting that Hawking paid off his bet with Thorne on whether Cygnus X-1 is truly a black hole; perhaps he felt that it is so close to being a black hole that it hardly matters, or perhaps he became persuaded that the inclusion of quantum physics throws doubt on the infinite-time calculation.

The fact that black holes don't yet really exist—at least "not yet" according to an outside frame—is a fine point, and normally not even mentioned to the nonexpert. But you might be able to win a bet using this "believe it or not" fact.

Another Lightspeed Loophole

In Chapter 5 I gave an example showing how acceleration of your proper frame at $1g$ could result in the distance between you and a distant object (measured in that accelerating frame) changing at a rate 2.6 times the speed of light. With the Lawrence Berkeley Laboratory electron accelerator BELLA, you could change the distance to Sirius in the electron's proper frame with an equivalent speed of 8.6 billion times lightspeed. You can do even better. You can change distances with infinite speed. Here's how.

Imagine that you and I are a few feet apart, in space, nothing else around. Assume that our proper frames are identical, so in that frame we are both at rest. Now, get a small primordial (completely formed) black hole, maybe one weighing only a few pounds. Plunk it right in between you and me. The gravitational attraction of the black hole is no bigger than for any other object with the same mass, so we don't feel any unusual forces. When the black hole is in place, the straight-line distance between you and me becomes infinite. You can see this on the black-hole diagram. The distance between us has changed. Yet our locations have not.

Have we "moved"? No. Has the distance between you and me changed? Yes. Enormously. Space is fluid and flexible. It can be compressed and stretched. An infinite concentration of space can be moved about easily, since it can be light in mass. That means distances between objects can change at arbitrarily rapid rates, even light-years per second, or faster. It is as if you are moving with super speed—although, in fact, you are not moving at all.

As I mentioned earlier, these concepts will prove important when we discuss modern cosmology in later chapters. In particular, they are the basis for the theory of inflation that is used to explain the puzzling paradox that the universe is remarkably uniform, even though it is so big that it never (apparently) had time to establish such uniformity. More on that later.

Wormholes

A wormhole is a hypothetical object, similar to a black hole, but instead of the curved space reaching down to an object with huge mass, it eventually spreads out and emerges at a different location. The simplest wormhole is very similar to two not quite black holes connected near the bottom. ("Not quite" means that you can fall in and shoot back out the other side in finite time.) For that to happen, you can imagine that space is folded, so that where the wormhole comes out is across the fold (see Figure 7.2). However, there is no need to imagine that. Remember that the depth to the bottom of the black hole is, from the outside reference frame, infinite in distance. So even though a wormhole isn't that deep, it could be deep enough to reach anywhere.

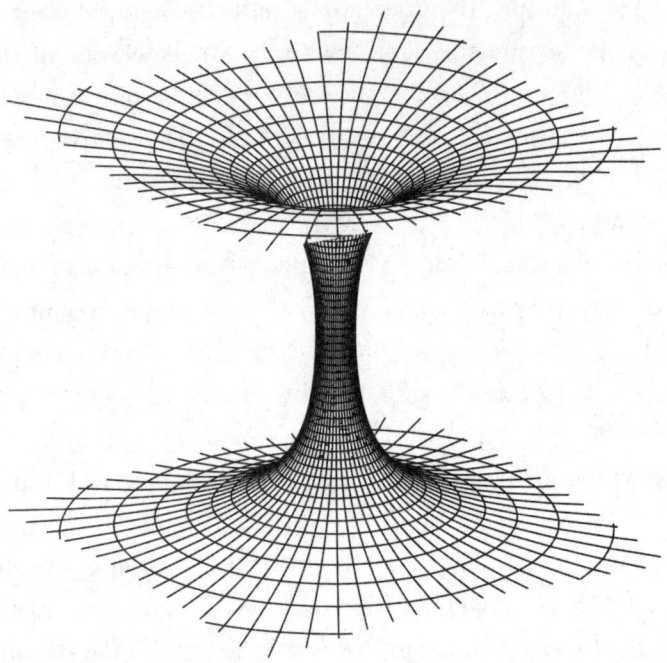

Figure 7.2. Conceptual depiction of a 2D wormhole. Two almost-black-holes connect two regions of space-time. Fall in one side, and pop out the other.

A problem with simple wormholes is that calculations show they are not stable. With no mass at the bottom to hold the curved space in place, the wormhole is expected to collapse faster than a person could shoot through it. We might be able to stabilize a wormhole (like stabilizing a coal mine by putting up columns), but current theory says that to do that, we would need something we haven't yet discovered, a kind of particle that has negative energy in its field. Such a field might be possible— at least, we can't rule it out—so science fiction is welcome to go ahead and assume that in the future we will have been able to create stable and useful wormholes.

Wormholes are the current science fiction orthodoxy for rapid travel covering distances of many light-years. Even the *Star Trek* term *warp drive,* also used in the *Doctor Who* series, suggests that the 4D space-time universe is bent into a fifth dimension, bringing distant objects close to each other. The same is true for the movie version of *Dune*, in which the Guild uses a special material known as *spice* to bend space. (In the novel, they simply cover distances faster than light, but the movie makes relativistic sense of that ability.)

Wormholes also fascinate science fiction fans because some physicists have argued that they would make travel backward in time possible. As we delve into the meaning of the flow of time, the meaning of *now* and of time travel, you'll see why I don't agree that passing through a wormhole could accomplish backward time travel.

IT IS AMAZING to me that, although we don't know why time flows, we can talk precisely about the relative flow of time in different locations, and that such flows happen at different rates. Time stretches and shrinks, depending on physics. The next step taken in physics also didn't explain the rate of flow of time, but it did address the simpler question of its direction: Why does time flow forward rather than backward?

PART II

BROKEN ARROW

8

An Arrow of Confusion

*Eddington proposes that increasing entropy explains
why time moves forward.*

> All the king's horses
> and all the king's men
> Couldn't put Humpty
> together again.
> — *Mother Goose*

Despite his enormous progress in understanding time, Einstein failed utterly in accounting for its most fundamental feature—that it moves. Time is not just a fourth spatial dimension. It is intrinsically different: it progresses. Moreover, the past is very different from the future; we know a great deal more about it. That special moment that gives this book its title, *now*, travels forward through time. Why? Could it move backward; could we discover how the time machine of H. G. Wells worked, and build one? We can change the future; at least our parents tell us we can. Why can't we change the past? Or can we?

Into this conundrum stepped Arthur Eddington. Eddington was a physicist, astronomer, philosopher, and popularizer of advanced science. He devised and conducted tough experiments, developed new theories, and got his name attached to important physics ideas. In 1919 he was asked about the claim that general relativity was so difficult that only three people in the world truly understood it. According to legend, his answer was: "Who's the third?"

Eddington made the first measurement of the deflection of starlight past the sun, one of the key tests of Einstein's curved space-time. Eddington did that difficult measurement in 1919, during a solar eclipse so that he could see the star unobscured by the brightness, and in doing so he made Einstein famous, and himself as well.*

Eddington thought deeply about physical phenomena. The *Eddington limit* is known to every astronomer and astrophysics student. It describes the balance that takes place in stars between the outward pressure of starlight and the inward pull of gravity, and it has proved key in the current understanding of not only giant stars but also exotic astronomical objects such as quasars.

Eddington knew that despite Einstein's great advances, there were still unexplained mysteries about time. In his 1928 book *The Nature of the Physical World*, Eddington wrote,

> The great thing about time is that it goes on. But this is an aspect of it which the physicist sometimes seems inclined to neglect.

In that book, Eddington didn't give any explanation for the meaning of *now*, nor did he provide insight on why time flows, but he did give us the most widely accepted explanation for the *direction* of time.

Eddington asked, Why does time go forward? Most people, when they first hear that question, regard it as stupid, like asking, Why do we remember the past rather than the future? These questions appear silly, until you ponder them. Physics doesn't seem to distinguish the past from the future; its laws seem to work equally well applied backward in time. If you know the past, you can use the laws of classical physics to predict the future. Yet it turns out that if you know the

---

* Some people now argue that Eddington's measurement agreed *too* well with Einstein's prediction—that his instruments were not good enough to be able to make such an accurate measurement—and that Eddington's results were not fully objective, but the most recent analysis gives Eddington full credit.

future, you can use those exact same laws to figure out what happened in the past. Eddington not only asked the silly question, but he gave an answer that fascinated physicists and still continues to intrigue today.

To explain his idea for the direction of time, Eddington asked us to imagine a picture of a series of events as a function of time. He called it a *space-time diagram* and referred to Hermann Minkowski (discussed in my Chapter 6). But let's use a less abstract version, one that retains all of the essential elements: a strip of film from a movie. (Think back to the olden days when movies were actually recorded as a series of photos on film rather than as bits stored in computer memory.) If you look at the individual frames, can you tell whether you are looking through the front of the film or the back? Making that determination is difficult unless some writing is shown, perhaps a road sign. If the sign read "ᴚᴉɓh ᴛxǝИ ᴛᴉxƎ ʎǝlǝʞɹǝ𐐒", then you'd know you were looking at the film through the wrong side. Large-scale nature is pretty much left-right symmetric (mountains, trees, and humans look just as real in a mirror), but culture isn't. Biology also breaks the symmetry; not only are most people right-handed, but so is common sucrose; its molecule has a right-handed twist.

Next question: Can you tell which way the movie should be played? What is the proper order of the frames? That's what Eddington referred to as the "arrow of time." If the movie showed planets moving around the sun, you probably wouldn't be able to tell the proper order. If the movie showed a close-up cartoon of atoms in a gas, bumping around, you probably couldn't tell. Yet for most movies, the arrow of time would be blatantly obvious. Run the film the wrong way and people would be walking backward. Pieces of pottery would jump up from the floor and reassemble into an unbroken teacup. Bullets would emerge from the murdered body and fly into the gun. Things sliding across surfaces would speed up from friction.

None of these peculiar actions violate physics laws. A broken egg could reassemble itself and then fly up on the table—if the molecular forces just happened to be organized in the right way. It's just very unlikely. Friction tends to slow things down, not to accelerate them. Heat flows from hot objects to cold ones, not the other way around. Crashes break objects; they don't pull them together. These observations can be given a precise formulation: it's called the *Second Law of Thermodynamics*. (The First Law says that energy can never be created or destroyed; of course, you have to include Einstein's mass energy, $E = mc^2$, when adding up the energy.)

The Second Law states that there is a quantity called *entropy* that, for a collection of objects, either stays constant with time or increases. Contrast that with energy, which always remains constant. Energy can shift from one thing to another, but the sum of the energy of all the objects doesn't change. Unlike the First Law, the Second Law is not absolute, but only probabilistic. Although it can be violated, the likelihood of a violation for large-scale collections of particles is negligibly small.

Entropy and time increase together. They are correlated. That was known. Eddington's new speculation was that entropy was responsible for the *arrow of time*, the fact that time moves forward rather than backward. He argued that the Second Law of Thermodynamics explains why we remember the past rather than the future.

Eddington's entropy-arrow linkage has such far-reaching implications for our understanding of reality, and maybe even consciousness, that some people think it is something all educated people should know about. C. P. Snow lamented, in his classic and highly influential 1959 book *The Two Cultures and the Scientific Revolution*, the fact that not all "educated" people knew of this great advance. He wrote,

> A good many times I have been present at gatherings of people who, by the standards of the traditional culture, are thought highly educated and who have with considerable gusto been expressing their incredulity at the illiteracy of scientists. Once or twice I have been provoked and have asked the company how many of them could describe the Second Law of Thermodynamics. The response

was cold: it was also negative. Yet I was asking something which is the scientific equivalent of: Have you read a work of Shakespeare's?

The Second Law of Thermodynamics compared to Shakespeare, by a serious scholar! I'm not sure I agree with Snow, even though his book was influential in my own life (it was freshman common reading at Columbia). Maybe Snow's "highly educated" people had never heard of the Second Law, but I would guess most of them knew enough physics to be able to comment intelligently about $E = mc^2$. Relativity would be a more appropriate analog to Shakespeare.

Eddington took the Second Law even further, elevating it to a mystical position at the pinnacle of science. He wrote,

> The second law of thermodynamics holds, I think, the supreme position among the laws of Nature. If someone points out to you that your pet theory of the universe is in disagreement with Maxwell's equations—then so much the worse for Maxwell's equations. If it is found to be contradicted by observation, well, these experimentalists do bungle things sometimes. But if your theory is found to be against the second law of thermodynamics I can give you no hope; there is nothing for it but to collapse in deepest humiliation.

This assertion sounds more like a religious tract than a statement by an eminent scientist. Yet Eddington's extravagant claim of the Second Law's "supreme position" has a simple basis. Deep down, the law is a statement that high-probability events are far more likely to occur than low-probability events. That's tautological, but being so makes it true. We'll talk about the probability interpretation soon, but to begin, let's make the Second Law a little less mysterious.

The heart of the Second Law of Thermodynamics is the concept of entropy. What is entropy?

9

Demystifying Entropy

Entropy sounds mystical, but it is also an engineering tool, with ordinary engineering units of calories per degree . . .

I am the Spirit that Denies!
And justly so: for all things, from the Void
Called forth, deserve to be destroyed . . .
— *Mephisto, in Goethe's* Faust

Physics has a way of giving obscure and abstract definitions to everyday quantities. Unless you graduated with a degree in physics, for example, you may be unfamiliar with the definition of *energy*, developed by Emmy Noether (see Chapter 3) that is taught in advanced courses:

Energy is the canonically conserved quantity corresponding to the absence of explicit time dependence in the Lagrangian.

Needless to say, that's not the way it is taught in high school, or even in most undergraduate physics courses, but it is very useful when new circumstances arise. For example, if you're Einstein and you've just derived some equations you call relativity, and you want to figure out how to redefine energy to be conserved with those new equations, you can just apply Noether's rule. (For more on this advanced understanding of energy, see Appendix 2.)

Some other physics quantities have equally abstract and mystifying definitions that prove to be useful for the expert but obscure to the non-physicist. One of these is the advanced definition of *entropy*. At its most abstract, the definition can be worded like this:

> Entropy is the logarithm of the number of quantum states accessible to a system.

That definition is about as easy to understand as Noether's definition of energy. Entropy appears to be arcane, abstract, and something beyond the ken of all but the most mathematically adept physicist-statisticians.

If that is your impression, then you might be surprised to learn that the entropy of a cup of coffee is about 700 calories per degree Celsius. The entropy of your body is about 100,000 calories per degree. With a little bit of physics and chemistry knowledge, and a chemistry handbook, you can figure out the entropy of common objects. If you are curious about this, look up "entropy of water" on the web.

Calories per degree? Those are the same units as heat capacity, taught in high school physics, giving the amount of heat you have to put into an object to raise its temperature. It doesn't really sound like the logarithm of the number of quantum states, does it? Nor does it sound like the "degree of disorder." Entropy may be a bit mysterious, but it is not mystical. It is everyday, and an essential tool in engineering.

The Motive Power of Fire

Just as computer technology drives the information revolution, the steam engine powered the industrial revolution. In the early 1700s, steam engines were huge, filling entire buildings, inefficient, yet still economical for pumping water out of deep mines. Rapid innovation took place in the midst of intense competition. By 1765, James Watt, who eventually had the unit of power named after him, had discovered ways to make the engines smaller and less wasteful of energy. In 1809, Robert Fulton was driving steamboats on six US rivers and the Chesapeake Bay. Finally, the engines became small enough to be mounted on a locomotive, transportation was transformed, and the American West was opened. The revolution has not stopped. Today's coal and natural-gas plants are advanced versions of steam engines, as are nuclear power plants, which use uranium in place of coal but still run on steam.

Most of the early development of the steam engine was empirical. James Watt, a Scottish instrument maker, recognized that energy was being wasted in steam engines by alternately heating and cooling the cylinder that drove the piston; he introduced a separate condenser that enormously improved the efficiency. But a theoretical understanding, a deeper way to optimize the approach without resorting to trial and error, the method we use today, awaited a young French military engineer named Sadi Carnot. He worked out the physics of steam engines in the early 1800s and reached some remarkable conclusions.

Carnot deduced that the working of the engine was not fundamentally dependent on the use of steam; steam engines were just one of a class of engines that extract "useful" mechanical energy from a hot gas. His analysis is used today for gasoline and diesel engines. Ideally, you'd like *all* of the heat energy turned into mechanical energy, but Carnot concluded that doing so is impossible. The fraction that can be so converted is called the *efficiency*. Carnot showed that keeping one side of the engine cold was just as important as keeping the other side hot; it is the ratio of hot to cold that sets the efficiency. In fact, the departure from perfect efficiency is just the ratio T_{cold}/T_{hot}, where the temperatures are measured on the absolute scale. If T_{cold} is low enough, or T_{hot} high enough, you can approach 100 percent efficiency.

A nuclear power plant today uses uranium to provide heat to make steam, and cooling water to turn it back to a liquid. The cooling towers, not the uranium fission reactors, are the structures that have become the icons of such plants, as seen in Figure 9.1. The nuclear fission takes place in the small domed building in the lower right of the photo. That's hardly prominent compared to the graceful cooling towers. These nuclear power plants are still based on Carnot's equations, using hot and cold in combination for maximum efficiency. So, even nuclear power plants are still just steam engines, as strange as that seems. Similarly, nuclear submarines run on steam.

Given a hot fluid (steam) and a cooling compartment, you still have to design a steam engine carefully to avoid wasting the heat energy. Carnot figured out the best way to do it, and today we call that optimum device a *Carnot engine*. We rank other engines by the percentage of Carnot efficiency they achieve. (Sometimes you may hear that a heat engine is 90 percent efficient; what that means is that it achieves 90 percent of the

Figure 9.1. A nuclear power plant. The energy is generated in the small domed building seen in the lower right. The large graceful tower provides cooling, necessary to attain high efficiency for electricity production. The "smoke" consists of a fog of (nonradioactive) water droplets produced from the lake.

Carnot efficiency.) The Carnot engine achieves high efficiency by reducing the excess entropy produced to zero. I'll define entropy in a moment, but its key feature for steam engines is that if you create entropy, you waste energy. Carnot didn't coin the term *entropy*; it came from one of his disciples, Rudolf Clausius, who took *en* and *y* from *energy* and stuck *trope*, meaning "transformation," in the middle. Clausius wrote in 1865,

> I propose to name the quantity S the entropy of the system, after the Greek word τροπη [*trope*], the transformation. I have deliberately chosen the word entropy to be as similar as possible to the word energy: the two quantities to be named by these words are so closely related in physical significance that a certain similarity in their names appears to be appropriate.

So, if you have confused energy and entropy, it is Clausius's fault.

Entropy from Heat Flow

In its original formulation, the entropy of an object was defined as zero when all heat is removed. To find the entropy of an object when it is warm, start at zero temperature (absolute scale) and gradually add heat, all the time keeping track of the rising temperature. A small increment of entropy is defined as the heat added, divided by the temperature. Add up all the little increments of entropy and you'll have the entropy of the warm object. That's how we measure the entropy of a cup of water. If you gradually cool the object, the entropy will decrease.

In general, cold objects have low entropy and hot objects have high entropy. In that sense entropy is similar to energy, but unlike energy, entropy is unlimited and readily produced. The total amount of energy in an isolated collection of objects doesn't change with time, although it can move from one object to another or transform from potential energy to kinetic energy or from mass to heat. That's the conservation of energy. Entropy, however, is not conserved. It can increase without limit.

In that sense it is similar to words; you can create as many new words as you like by talking. Words are not conserved. (Richard Feynman's dad used to tease his son about this; he would tell little Richard to keep quiet or he might run out of words and no longer be able to speak.) Entropy is the same way. The universe is constantly creating more.

Entropy can grow with time, even if you are doing nothing. It's easy to create entropy. Take a cup of hot coffee and let it stand in a cool room. As heat leaves the coffee, the entropy of the coffee decreases (negative heat flow), but the entropy of the room increases more than enough to make up for it.* So, by just putting the coffee aside to cool, you are responsible for purposefully increasing the entropy of the universe, something that can never be undone.

The Second Law of Thermodynamics says that for any isolated system, the entropy will either remain constant or increase. The words *isolated system* are included because local entropy can decrease (for example, in the cooling coffee cup), but only if the entropy elsewhere (the room) increases. The Second Law of Thermodynamics does allow entropy to remain constant; the entropy of an object sitting at equilibrium doesn't change. A perfect Carnot engine operates without increasing the entropy of the universe; that's why it's so efficient.

Entropy has all sorts of practical uses. Chemists use tables of entropy of common chemicals to determine which chemical reactions will take place and which ones won't. Unless the calculated entropy of your initial chemicals is less than that of the reaction products, your reaction won't go. It's the Law.

When our political and ecological leaders urge us to "save energy," what they really mean is for us to generate as little additional entropy as possible. Entropy generation implies that energy has been "wasted"; it has flowed from hot to cold without producing useful piston-pushing work.

---

* The entropy loss of the hot coffee is $-\text{heat}/T_{cup}$. The entropy gain of the room is $+\text{heat}/T_{room}$. The heat values are the same (except for sign), but since T_{room} is smaller than T_{cup}, the entropy loss of the cup is less than the entropy gain of the room.

No real engine actually achieves Carnot efficiency, so saving energy also means getting by with as little useful work energy as possible. In the end, even useful work energy winds up as heat, and that, too, increases the entropy of the universe.

Entropy of Mixing

Letting heat flow is not the only way to create entropy. For example, you can take carbon dioxide from a coal power plant and let it mix into the atmosphere. The resulting "entropy of mixing" can be easily calculated using rules figured out by Carnot, Clausius, and their successors. Their formulas are standard fare in junior-level physics courses. Likewise, when you put chocolate syrup in your milk, you mix two fluids, and without additional energy use, you can't unmix them. This entropy of mixing will make more sense when we discuss, in the next chapter, the relationship between entropy and confusion.

Here is a practical example. Let's say you want to desalinate ocean water. Such water is a mixture of salt and water, and it has an entropy of mixing. If you desalinate the water, you eliminate the entropy of mixing. The Second Law says you can do that only by increasing the entropy somewhere else—for example, by using heat flow to push a piston that puts pressure on the salt water, driving it through a membrane that separates the two components. A calculation shows that a minimum amount of energy must be expended to desalinate; it turns out to be about 1 kilowatt-hour of energy for every cubic meter of seawater that you purify.

That number has practical value. I once reviewed a business proposal for a new desalination method; the first thing I checked was whether its extraordinary claims violated the Second Law of Thermodynamics. They did, so I advised the investor to stay away. The inventor was violating the Law.

Entropy calculations can not only tell us that some claims are fraudulent, but also set a goal for what is achievable. If we say that electrical energy costs 10 cents per kilowatt-hour, then the cost of the 1 kilowatt-hour required to desalinate a cubic meter of seawater will be a minimum of 10 cents. That translates to about $100 per acre-foot (roughly the amount of water a family of five uses in a year). Currently, desalination plants are not nearly so cheap; they deliver freshwater for about $2,000 per acre-foot. So, the Law allows a potential cost reduction by as much as a factor of 20. Traditional farm-water prices in California have been typically $6–$40 per acre-foot, making desalination unprofitable, but during the 2015 drought, some farmers paid as much as $2,000 per acre-foot. That cost made desalination competitive. (Of course, investing in a desalination plant is still risky, since the price of water drops when a drought ends.)

One way to reduce the cost of desalination is to use energy that is cheaper than electricity. For example, you can use direct sunlight for the heating that drives the plants. Such plants now exist in the Middle East. Incidentally, you can also use sunlight to cool. Try to guess who holds that patent! Here's the amazing answer: the US patent for the sunlight-powered refrigerator, number 1781541, is held by Albert Einstein and physicist Leo Szilard (who also patented the atom bomb). You might enjoy looking it up on the web. It is also a sufficiently surprising yet obscure fact that you might be able to win a bet about it.

Entropy calculations also have implications for the removal of carbon dioxide from the atmosphere, for those of us concerned about global warming. Dump carbon dioxide into the air, and a thousand years later almost a quarter of it will still be there. Although in principle it can be removed, it is mixed in a huge atmosphere, so the entropy of mixing is enormous. Extracting the carbon dioxide means you have to produce entropy elsewhere (typically in the form of heat), and that takes a great deal of energy. It would be much cheaper to capture the carbon dioxide before it mixes with the atmosphere. Or we could just leave the carbon in the ground.

Enough about the practical use of entropy calculations. It is the abstract mystical interpretation of entropy that relates it to the behavior of time.

10

Mystifying Entropy

The deeper meaning of entropy *is one of the most fascinating discoveries in the history of physics . . .*

The most amazing aspect of entropy was hiding deeply buried underneath its engineering facade. The simple concept of heat flow and temperature was concealing its roots in the quantum world. It was slowly uncovered as scientists in the 1800s sought to create and understand a new discipline called *statistical physics*, based on unverified assumptions about the existence of atoms and molecules. Mysteries and paradoxes in statistical physics led scientists to the discovery of quantum physics—and Eddington to the idea that the flow of time was driven by entropy increase.

The Physics of Umpteen Things

Physics is very good at predicting the behavior of one or two atoms. It can also handle one or two planets. The toughest realm is when a handful of objects interact. It turns out to be very difficult to predict whether a triple-star system is stable; we know the equations, but math has been unable to "solve" them—that is, to write the solution in terms of the common scientific functions, like exponentials and cosines, that can readily be evaluated. We can simulate the motion of a handful of stars on a computer, and that's typically what is done. Alas, the behavior of a triple-star system tends to be chaotic, and highly precise values for the initial positions and velocities are needed just to make rough estimates of the future. As a result, in astronomy it is often uncertain whether or not a particular star system is stable or whether at some time in the future one of the stars will fly off to infinity.

Remarkably, as the number of objects increases, the physics gets easier. That's because for many important problems, all we really want to know are the averages, and with large numbers of particles (there are 10^{23} molecules in a gallon of air), the averages can be accurately determined. We can even calculate the average departures from average.

Before statistical physics was developed, simple gas laws were discovered empirically. Back in 1676, Irish chemist and theologian Robert Boyle showed in a series of experiments that the pressure of a given amount of air is inversely proportional to its volume. Compress it to half of its volume and the pressure doubles (if you keep the temperature constant). In the 1800s, statistical physics accounted for this result by postulating that the gas consisted of a huge number of microscopic atoms and that pressure was simply the average result of a huge number of tiny collisions of these atoms with the walls.

The explanation of the behavior of gases in terms of atoms was one of the early great "unifications" of physics. Prior to the atomic theory, the behavior of gases was not related to Newton's laws (such as $F = ma$). Heat was thought to be a separate fluid, *caloric*, that intermingled with the gas. But the statistical physicists said that heat simply represented the energy of the individual atoms; rapidly bouncing atoms were "hot" and slow ones were "cold," and the temperature (in degrees absolute) was the average kinetic energy per atom.

Once again, in walked Einstein to play a pivotal role. In 1905, the same year that he showed $E = mc^2$, he was looking at the possibility of testing the atomic theory by calculating the effect that atoms would have on small dust grains. When beginning work on this question, he discovered that the effect may have already been observed, by botanist Robert Brown in 1827, and was known as *Brownian motion*. Viewing through a powerful microscope, Brown observed that tiny pollen grains wiggled and squirmed and moved along as if they were trying to swim. The favored speculation at the time was that such tiny specks had incipient life, that they were proto-paramecia or something like that, showing an intrinsic primordial life force.

No. Einstein showed that the motion was just what you would expect if the molecules of water bombarding the pollen from opposite sides did not exactly average out. Small shoves on one side would, from time to time, be greater than on the other side, and the dust would jump in response. Although on average a particle would stay in the same place, he calculated the departure from the average. The particle did undergo net motion, but not because it was swimming; it was a random walk, also known colorfully as a "drunkard's walk." Take lots of steps in random directions and you will wander away from your starting point; your distance from it will increase, on average, by the step size multiplied by the square root of the number of steps. After initial experimental tests seemed to indicate that Einstein was wrong about Brownian motion, careful measurements made by Jean Perrin in 1908 confirmed Einstein's predictions and led directly to the widespread acceptance of the existence of atoms and molecules—and to the acceptance of statistical physics.

I find it a remarkable fact about physics that even though we knew a lot about electricity, magnetism, mass, and acceleration in the late 1800s, it was not until the work of Einstein and Perrin in 1905–8 that the scientific community as a whole accepted the existence of atoms and molecules.

When I was a teenager in the 1950s, I read the book *One Two Three . . . Infinity* by George Gamow. It included a photograph of a "hexamethylbenzene molecule." It showed twelve black spots in a regular hexagonal pattern; I assumed the spots were individual atoms. (They weren't; I now know they were clusters of atoms.) I found that photo very exciting.

Atoms had been photographed! These days, photographs of atoms are common, but back in 1989 IBM made the news when it arranged thirty-five xenon atoms on a surface to form the letters "IBM" and made an image of that, all with a new device called a *scanning tunneling microscope*. Atoms are no longer hypotheses, but in Einstein's day, that's all they were.

Einstein's explanation of Brownian motion would have been considered the greatest physics advance of the year, even of the new century, except for the fact that in the same year, Einstein wrote three other great papers: two on relativity, and one postulating the quantum nature of light. This latter paper, on the "photoelectric effect," is what was used as an excuse to award him a Nobel Prize. This astonishingly productive year for Einstein has been called by physicists the *annus mirabilis*, the "miracle year."

What Is Entropy, Really?

Statistical physics showed that pressure comes from bouncing particles and that temperature is the kinetic energy per particle. Entropy had a more subtle and remarkable explanation, worked out by physicist-philosopher Ludwig Boltzmann nearly four decades before Einstein's Brownian-motion work. Boltzmann spent much effort defending his theories of statistical physics. He suffered from what we now call bipolar disorder, and in 1906, during a fit of depression, he hanged himself, just three years before Perrin's experiments convinced the physics world that the basic assumptions of his work were right.

Boltzmann had shown that the entropy of a material is related to the number of different ways that molecules can fill the volume to make the observed macroscopic state. That number is called the *multiplicity*. Imagine a gallon of air, with 10^{23} different molecules. In one state, all the molecules might be clustered together in one corner. There is only one way to achieve that arrangement, so the multiplicity of that state is 1. In another state, the molecules might be spread out, with an equal number in every cubic centimeter. The multiplicity of that state is huge, since we could put the first molecule in any of the 3,785 different cubic centimeters in that gallon, the second in any other, and so on, making sure not to overfill any particular cubic centimeter. Since the number of molecules of air in a gallon is large, 10^{23}, the multiplicity, the number of different ways to fill those cubic centimeters, is gigantic, but calculable. (We'll get to some actual numbers in a moment.)

Boltzmann proposed that the multiplicity of a state gives the probability of that state. So, having the molecules fill the space evenly was enormously more likely. In calculating the multiplicity, Boltzmann also included the number of different ways the particles could share the available energy.

Boltzmann discovered that that this approach was the key toward understanding the deeper meaning of entropy. Once he had calculated the multiplicity W of a state, he found that the logarithm of that number gave a number proportional to the entropy! That was an astonishing discovery. Previously, entropy had been an engineering term used to minimize wasted heat. Boltzmann showed that it was a fundamental quantity rooted in the abstract mathematics of statistical physics. Here's his equation:

$$\text{Entropy} = k \log W$$

The value of k was chosen to convert $\log W$, a pure number, to engineers' entropy, measured in calories per degree or joules per degree. Today, k is called *Boltzmann's constant*. (I used the same letter k in Einstein's equation for general relativity, but this is a different number.) It is so useful that every physics student learns to remember its value.* Boltzmann was so proud of this achievement that he asked for this equation to be carved on his tombstone, and it was, as shown in Figure 10.1.

---

* In physics units and using the natural logarithm, k is 1.38×10^{-23} joules per degree Kelvin. In the units of this discussion and using the logarithm base 10, k is 7.9×10^{-24} calories per degree Kelvin.

Figure 10.1. Ludwig Boltzmann's tombstone,
showing his equation for entropy near the top.

The number *googol* was made up by nine-year-old Milton Sirotta, when his mathematician uncle Edward Kasner asked him to come up with a name for 1 followed by as many zeros as he could write. Later they decided that a googol would be 1 followed by a hundred zeros. We can write that as 1 googol = 10^{100}. (The company Google was named after a misspelling of this word by Sean Anderson, a friend of founder Larry Page.) The number of atoms in the universe is estimated to be 10^{78}, less than a googol by a factor of 1 followed by twenty-two zeros. But the multiplicity of a container of gas, the number of ways to fill it, is typically 1 followed by 10^{25} zeros. That's $10^{10^{25}}$, immensely greater than a mere googol. However, it is less than a googolplex.

What is a *googolplex*? That humongous number is defined as 1 followed by a googol of zeros. (It was also the original name for Google suggested by Anderson.) It can be written as $10^{10^{100}}$. It is so huge that many people think it has no relation to reality. It is greater than the number of cubic millimeters in the known universe. But it does come up in statistical physics when we calculate the entropy of that universe, estimated by Chas Egan and Charles Lineweaver to be $3 \times 10^{104} k$. Remember, this huge number is the logarithm of the multiplicity W. W is very much larger.* The number of different ways you could rearrange all the things that make up the universe without changing our current state (same stars and other entities)—that is, W for the universe—is larger than a googolplex, much larger, about $10^{10^{104}}$. That means that the multiplicity of the universe is bigger than a googolplex by a factor of 1 followed by 10,000 zeros.

The Tyranny of Entropy

How will real molecules distribute themselves in a real container? How will they divide up the available energy? The key insight of Boltzmann was that the state with the largest multiplicity W dominates. Higher entropy wins, and it wins big, because the relative probabilities are not determined by $\log W$ but by W itself, and W is much bigger than $\log W$.

The results of statistical physics require the assumption that the probability of any particular state depends on the number of different ways it could come about, the multiplicity. This assumption is not self-evident. It is called the *ergodic hypothesis*. In fact, it is not strictly true. If you had two containers, one filled with gas and the other one empty, the state with the highest entropy would have half of the gas in each container. But if the containers were not connected, there would be no way for the gas to move from one to the other. The highest-probability state is unreachable.

This may sound like a trivial caveat, but it will prove to be very

---

* The logarithm of a number, to base 10, is approximately the number of digits in the number. For example, the logarithm of 1,000,000 is 6.

important to an understanding of time. It forces us to redefine entropy like this: the logarithm not of the number of ways of filling the boxes, but of the *accessible* ways of doing it. When you do your counting, don't count the ways of filling the boxes that violate some other law of physics, such as molecules passing through walls. In the rest of this book, the multiplicity W will represent the number of accessible ways of filling the boxes.

Humans may not be able to stop entropy increase, but we can exercise some control over the accessible states. I'll argue later in this book that such guidance is the key value of human free choice. We can't lower the entropy of the universe, but we can choose whether or not to connect the two gas containers. If we don't connect them, the entropy of the universe is lower than it otherwise would be.

We can also manipulate local entropy, lowering it if we wish. That's what an air conditioner does. It cools the indoor air, lowering the entropy in the house, and ejects heat out the back. The increased entropy of the slightly warmed outdoors is larger than the amount of entropy lost indoors. Thus, running the device cools us and lowers our own entropy, but raises the net entropy of the universe.

Life represents a local decrease in entropy. A plant takes rare and dispersed carbon out of the air, combines it with water concentrated from the soil, and, using the energy of sunlight, creates complex molecules of starches and arranges them into highly organized structures. The entropy of the molecules making up the plant is lowered, but the net entropy, mostly heat thrown off into the atmosphere, increases.

Entropy Is Confusion

Entropy is often said to measure the state of confusion, the state of disorder. The low-entropy state of a gas, with all the molecules in one corner, is highly organized. The high-entropy state, with molecules spread out, is disorganized. *High entropy* refers to a state that is very likely to occur from random processes. *Low entropy* means an organization that is improbable. A highly organized state is, almost by definition, one that you don't find resulting from random natural processes.

In principle, when you do something to a system, such as running an ideal Carnot heat engine to extract useful mechanical work from hot gas, the total entropy could remain constant. But such a perfect engine has never been built. In practice, entropy always increases, meaning that increased disorder is inevitable. Heat flow from a hot object to a cold object increases entropy. The universe is losing its net organization and slowly but surely becoming randomized.

Smash a teacup and you increase the entropy of the molecules. Broken, they are closer to the natural random state from which they came. Smash the cup completely, vaporize it into its individual molecules, eject the molecules into space, and let them spread about, and you lose all order and maximize the entropy. We reduce local entropy, at the expense of the rest of the universe, by building a teacup. Most of what we consider civilization is based on local decreases in entropy.

Entropy and Quantum Physics

Statistical physics led in a very surprising direction, to the discovery of quantum physics. Heat an object to several thousand degrees Fahrenheit and it will emit visible light; it will glow red-hot. Statistical physics attributed this radiation to the vibrating molecules in the object, since shaking electric charge emits light waves. The problem was that the calculation using statistical physics showed that this radiation would carry infinite power. Because the infinity came from short-wavelength (ultraviolet) light, the problem was termed the *ultraviolet catastrophe*, and it was a deep embarrassment and failure of statistical physics.

A German physicist named Max Planck proposed a strange and unphysical solution. He found an equation that would account for the actual observations. We now call that the *Planck formula*. It was math, not physics. He then searched for an assumption, a new physics principle that, if true, could be used to derive the equation. He found one: the idea that atoms could emit light only in *quantized* amounts. This astonishing idea was the founding principle of quantum physics.

Planck had to assume that, when an atom emits light at frequency f, the energy of that light must be a multiple of a basic energy unit, which he wrote as

$$E = hf$$

He picked the number h so that the observed radiation of hot objects would match his formula. We now call this number *Planck's constant*, and it is one of the most famous numbers in physics. Physicists often say that any formula that does not contain h is a "classical physics" equation, and any one that contains h is a "quantum physics" result.

Planck's assumption was ad hoc and arbitrary. His equation matched the data, but his postulate of quantized light emission had no justification in physics. That was in 1901. Four years later, Einstein recognized that a somewhat different interpretation of Planck's law could be used to explain a completely different mystery, the photoelectric effect. The photoelectric effect is the basis of present-day solar cells and digital cameras. It had been discovered in 1887 by Heinrich Hertz (the same German physicist who discovered radio waves and after whom the unit hertz, as in 60-hertz electricity, is named).

Hertz discovered that light hitting a surface knocks off electrons. But he found that the energy of the emitted electron depends on the color (that is, the frequency) of the light, and not on its intensity. That discovery was completely mysterious. Upon increasing the intensity of the light, Hertz did not get higher-energy electrons, but he did get more of them. That observation, too, did not make sense if light was an electromagnetic wave.

Einstein realized he could explain Hertz's photoelectric effect if he assumed that light itself was quantized. (Planck had assumed it was the emitting atom that was quantized.) Einstein called these bundles light *quanta*; later scientists called them *photons*. Einstein, in effect, discovered the photon; at least, he was the first person to recognize its existence. Each photon knocks out one electron. It gives its energy hf to that electron, so the electron energy depends on the frequency of the light. More intense light just means more photons, and thus more electrons ejected. It was Einstein's explanation of the photoelectric effect that earned him a Nobel Prize in 1921.

It is ironic that the quantum explanation of the photoelectric effect also established Einstein as one of the founders of quantum theory. The irony is that he never accepted that theory, at least not the version of it that grew to dominate physics.

ENTROPY INCREASES. Time progresses. Are they just correlated, or are they causally related? Arthur Eddington argued that they were linked, but not in the obvious way. Entropy was not merely increasing with time, as the statistical physicists had deduced. Eddington argued that it was the other way around: Entropy was the driver. Entropy was the reason that time moved forward.

11

Time Explained

Eddington explains how entropy sets the arrow of time.

Problems worthy of attack
prove their worth by hitting back.
— *Piet Hein*

The Only Thing We Have

Ask a random physicist, "What makes time move on?" I don't know how many random physicists you know personally, but I know a goodly number, and I've tried that question on many of them. The answer I get usually is something like this: "Probably entropy." Then the physicist qualifies the answer: "I'm not sure that's true, but it seems to be the only thing we have."

Perhaps the most interesting part of the answer is that the random physicist has actually thought about the problem. This was the sort of question that, a century ago, you would have asked a philosopher, not a scientist. Look at what Schopenhauer said, or Nietzsche, or Kant (although he was also a scientist), and you'll find that, indeed, they all spoke to the subject. Before the Enlightenment, you probably would have asked a priest or a theologian, such as Augustine or Ockham. But thanks to Einstein, such issues became part of physics. Today, how could you even begin to address such a question if you didn't understand relativity, if you didn't understand Einstein's great advances in the behavior of time and space?

Arthur Eddington's 1928 book *The Nature of the Physical World*, in which he argued that the arrow of time was set by entropy, was not written in a highly technical style (despite Eddington's mastery of advanced mathematics); nor was it written as if aimed at experts, although it was. It is a book you can enjoy reading today (and because it's beyond copyright, it is available for free on the Internet). It wasn't quite a regression to childhood in Einstein's manner, but it certainly reflected the thesis that on the issue of time, simplicity is called for.

Eddington argued that only one law of physics had an arrow of time: the Second Law of Thermodynamics. Every other physical theory— classical mechanics, electricity and magnetism, and even the still-evolving field of quantum physics, appeared unable to distinguish past from future. Planets could move backward in their orbits, following exactly the same rules. An antenna for emitting radio waves could just as readily be an antenna for receiving those waves. Atoms emit light, but they also absorb light; the same equations described both. Run the

Figure 11.1. Arthur Eddington, in 1928.

movie backward and you violate no physics laws—none, that is, except the Second Law of Thermodynamics. None, except the law that says entropy will always increase with time.

Today, compelling evidence suggests that an arrow of time is built into the fundamental behavior of at least one additional realm of physics. It is in the physics of radioactive decay, historically referred to as the "weak interaction," and there is now evidence that "time reversal symmetry" is violated in some such decays. But this fact has not changed the minds of physicists about the arrow of time; they stick with the entropy explanation. I'll come back to this after I discuss Eddington's entropy arrow.

Movies Played Backward

Earlier I asked you to imagine a movie strip with a teacup falling from a table. You could tell which way the movie should be played because of the fact that teacups could, but don't, jump from the floor up to a tabletop and reassemble themselves. Small molecular forces, if they all just happened to align in the same direction, could accomplish that, but the probability of that happening is vanishingly small. So even if you weren't told which way to run the movie, the arrow of time is evident. The teacup is a favorite example, but undoubtedly you can think of many more. Stars burn out. Petroleum reserves become depleted. Mountains erode. We die and decay. The increase in entropy is inescapable.

Suppose you were given complete God-like knowledge of the universe for two instants of time and asked to figure out which instant came first. How would you do it? The simple answer: calculate the entropy of the two snapshots. Whichever one has lower entropy came first. Physicists found entropy to provide a pretty convincing arrow.

Primary and Secondary Laws of Physics

The Second Law of Thermodynamics, the law stating that entropy increases, is a rather odd law. It doesn't really add anything to physics other than the statement that behavior that is more probable is more likely to happen. Why does that qualify as a *law* of physics? Isn't that both self-evident and trivial, a tautology? And if the equations of mechanics and electricity and magnetism—the real physics—don't give a direction of time, why should a trivial law that is based on them do the job?

Eddington was well aware of this paradox. In fact, he made a distinction in the laws of physics between primary and secondary laws. Entropy was definitely a secondary law, derived from the others and not really standing on its own.

Let's make the paradox stronger. Let's assume the validity of classical physics, the physics on which the Second Law was based. In that physics, if you knew the position and motion of every particle (ignoring the uncertainty principle of quantum physics) couldn't you, at least in principle, predict the future exactly? There would be no need for probabilistic calculations, for the law of chance. So, how could the fundamental laws, lacking an arrow, give rise to a secondary law that does have that arrow?

The answer is that the present universe, for reasons that Eddington couldn't initially fathom, is highly organized. We have low entropy. When you take a gas that is confined to a corner of a box and let it spread throughout the box, you get an enormous increase in entropy. The matter in the universe is mostly compact, just like the gas stuck in the corner. Most of the visible mass is found in stars, with a little bit in planets, surrounded by mostly empty space. (I'm ignoring dark matter for now, as something that was unknown to Eddington.) There is lots of empty space that we can fill to increase entropy. Put another way, the organization we see around us is highly unlikely. Thanks to the fact that the universe is surprisingly well organized, and highly likely to move toward greater disorganization, time moves forward.

If you believe that the universe is infinitely old, then it has had an infinite time to evolve, an infinite time for entropy to increase, and you might think that the maximum in entropy would have been reached a long time ago. Why hasn't it?

Why Is the Universe So Improbable?

Some people think the current state of the universe, with its relatively high organization and low entropy (compared to what it could be), implies the existence of God. Eddington put it more elegantly. He said (in his 1928 book),

> The direction of time's arrow could only be determined by that incongruous mixture of theology and statistics known as the second law of thermodynamics; or, to be more explicit, the direction of the arrow could be determined by statistical rules, but its significance as a governing fact "making sense of the world" could only be deduced on teleological assumptions.

Just in case you are not up to date on your philosophy terms, here is the *Oxford English Dictionary* definition of *teleological*:

> Of, pertaining to, or relating to ends or final causes; dealing with design or purpose, especially in natural phenomena.

Time moves forward because our current state is so highly improbable. We have big concentrations of mass, lots of empty space, nonuniform temperatures. Therefore, heat can flow, objects can break, and mass can disperse through empty space. It didn't necessarily require God to get the universe organized, but the universe is indeed organized.

Of course, if it was God who created a universe with low entropy, it could well have been a Spinoza version—a God who was the creator and then just let the world evolve. This approach is called Deism. It's not clear that such a God cares about being worshipped, or that he is even worthy of it. Many theologians consider Deism to be a variant of atheism—a way of saying you believe in God even though you don't really.

Indeed, just as Eddington was writing his book, another fantastic discovery was unfolding on the other side of the world, in Pasadena, California, by astronomer Edwin Hubble. His discovery led to a theory that gave a physics explanation for the high organization required for time to have an arrow. It explained that the universe is organized because it is still young, relatively speaking. The name of the theory to explain this, the name we use today, was given to it by astronomer Fred Hoyle, who was trying to ridicule it. He called it the *Big Bang*.

<h1>12</h1>

<h1>Our Unlikely Universe</h1>

For entropy to increase, as required by Eddington, the present universe must have low entropy. How could that have come about?

If you gaze for long into an abyss, the abyss will also gaze into you.
— *Friedrich Nietzsche*

In 1929, Edwin Hubble made a discovery that seemed to set science backward by four hundred years. Yes, the Earth still did orbit the sun, as Copernicus had said, but on the larger scale, Ptolemy appeared to be right. Our galaxy, the assemblage of stars that surrounds us, the Milky Way, appeared to be at the center of the universe.

To understand what Hubble discovered, let's first review what he was looking at. Hubble was studying galaxies, huge collections of stars similar to our Milky Way. Figure 12.1 shows what the Milky Way probably looks like, as we would see if we could get outside of it to take a photo. It is actually an image of the nearby galaxy in the constellation of Andromeda.

In this photo, about a trillion stars are swirling around in almost circular paths. If this truly were the Milky Way, the sun would be one of them, about halfway out from the center. Essentially all stars we see at night are within our own galaxy. If you look up on a clear winter night, away from the lights of the city, you can see a small smudge almost overhead, roughly the angular size of the full moon. That's the Andromeda Galaxy shown in the photo.

Figure 12.1. What our galaxy, the Milky Way, would look like from the outside. (This is actually a photo of the Andromeda Galaxy.)

But trillions of such galaxies are visible in the night sky if we peer out between the nearby stars with our best telescopes. There are more such galaxies than there are stars in our own galaxy. The individual stars you see in the photo of the Andromeda Galaxy are not in the background but in the foreground. We are peering through them to see the Andromeda Galaxy behind.

That bright spot you see in the center of the galaxy contains several billion stars surrounding what astronomers believe is a supermassive black hole, itself containing the equivalent of about 4 million stars.

Prior to Hubble's work, most astronomers thought such galaxies were actually nearby puffs of gas, with the nearby stars in the background. In 1926, Hubble found convincing proof that they were actually vast collections of stars, far more distant than the stars in the named constellations. Then came his astounding discovery that the twenty-four galaxies he was studying were all moving away from ours in a remarkable pattern. The more distant the galaxy, the faster it was receding. It was as if there had been a big explosion, right at our location, and the chunks that moved the fastest were now the farthest away.

Hubble estimated the time of the explosion to be about 4 billion years ago, but his distance calibration was off. Using the same data but correcting that error, we now estimate the time of the explosion to be about 14 billion years ago. Moreover, we now know that Hubble's discovery is true, not just for twenty-four nearby galaxies, but for the hundreds of billions of other galaxies that are visible with the instrument now named, in his honor, the "Hubble Space Telescope."

It was one of the most, if not the most, important experimental discoveries of the twentieth century, and that was a century with many great discoveries. It certainly was the discovery most relevant to teleology since Nicolaus Copernicus had concluded four hundred years earlier that the Earth orbits the sun. Hubble seemed to be putting the Milky Way at the center.

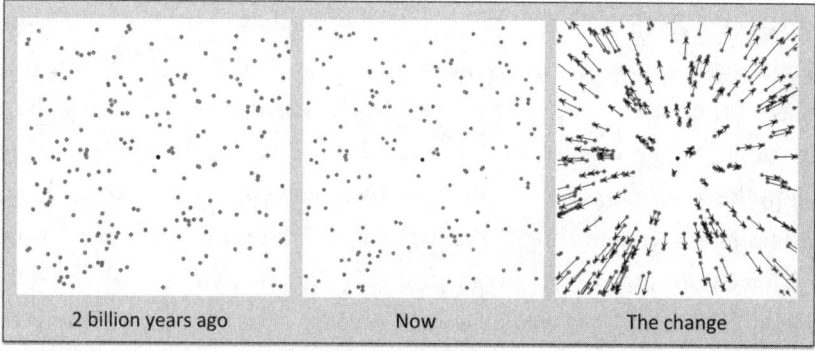

| 2 billion years ago | Now | The change |

Figure 12.2. The Hubble expansion, for randomly spaced galaxies in the past (left), as they are now (middle), and at both times (right), with arrows showing the change in position of each galaxy. Look closely at the center diagram and you'll see it's the same pattern as on the left, but slightly expanded. The galaxies appear to be moving away from the one in the center, but if we sat on any galaxy, they would appear to be moving away from that one.

But that interpretation is incorrect. Hubble's discovery did not put us back at the center, as Hubble knew. Put yourself in the proper frame of one of those galaxies flying out. They are all getting farther and farther apart. In your proper frame, all objects are moving away from you. It doesn't matter which galaxy you ride on; Hubble's law is the same for each.

This remarkable feature of Hubble's law is easiest to visualize with a raisin bread metaphor. Imagine you are a raisin in a loaf that is baking and expanding. All your neighbor raisins are getting farther and farther away. Raisins twice as far away are moving away from you twice as fast. You might draw the conclusion that you are in the center, unless you can see the crust, but you probably aren't. You'll get the same law from the frame of any raisin. So, although the public (mistakenly) thought the discovery put the Earth at the center, Hubble was quick to explain that it didn't.

No Crust Needed

Yet another interpretation of the expansion was even more fantastic. This idea was proposed two years before Hubble's discovery, by George Lemaître, a Belgian priest and professor of physics at the Catholic University of Louvain. Lemaître had proposed a model, based on general relativity, in which he described the early universe as a "Cosmic Egg exploding at the moment of the creation." He also referred to it as his "hypothesis of the primeval atom." Some people think he, not Hubble, should be given credit for the expanding universe, but Lemaître's work was based on some of Hubble's preliminary results and was published in a Belgian journal that was little read outside of that country. Lemaître has been called "the greatest scientist you have never heard of."

Lemaître was a student of general relativity, and he applied it to the universe as a whole. Hubble's discoveries led him to conclude that the universe was indeed expanding. By Lemaître's reckoning, however, what was exploding was not matter within a fixed space, but space itself. This concept was readily accommodated by Einstein's equations.

Figure 12.3. George Lemaître, theorist for the expanding universe.

Einstein had assumed that the universe was static, and he had added a term to his equations called the *cosmological constant* that provided a repulsive force to overcome the mutual gravity that would otherwise cause the universe to collapse. He thought Lemaître's idea of an expanding universe was ridiculous. He told Lemaître, "Your calculations are correct, but your physics is atrocious."

After Hubble's discovery, Lemaître was suddenly famous. On January 19, 1931, headlines in the *New York Times* declared, "Lemaître Suggests One Single, Great Atom, Embracing All Energy, Started the Universe." Einstein not only withdrew his cosmological constant, but regretted ever putting it in. George Gamow said that Einstein described his inclusion of that term as "the biggest blunder of my life." (There is a great irony in this statement, since today we believe the cosmological constant is, in fact, present and important in cosmology. I'll have more to say on this when I talk about dark energy.)

A newspaper in 1933 reported that after a lecture by Lemaître at Princeton, Einstein stood up and said, "This is the most beautiful and satisfactory explanation of creation to which I have ever listened." He had clearly changed his mind about Lemaître's "atrocious" physics. Lemaître also proposed that cosmic rays, radiation from space that had been discovered in 1912, could be artifacts left over from the explosion. On that issue, Lemaître was wrong. There was a remnant, but it turned out to be microwaves, not cosmic rays. People tend to forget the wrong theories of theorists; they remember only those that were right. Alas, the same is not true for experimentalists.

In Lemaître's math, every galaxy was sitting at a fixed point in space. Hubble's law came about not from galaxy movement, but from the expansion of space between the galaxies. It was yet another example of Einstein's equations allowing a rubbery nature for space. We have already seen the flexibility of space in several aspects of relativity (Chapter 2), including in the pole-in-the-barn paradox (Chapter 4), and in the two lightspeed loopholes (Chapter 5).

The Lemaître model is what we use today, although it is sometimes referred to as the Friedman-Lemaître-Robertson-Walker (FLRW) model after other cosmologists who developed similar approaches. This is the model that led to the fulfilled predictions about the nature of the very distant universe. Cosmologists soon coined a term, the *cosmological principle*, that summarized this approximation: the universe is the same everywhere else as it is here.

About 14 billion years ago, matter was tightly compressed, and then space exploded. The matter, in fixed space coordinates, went along for the ride—not moving, but getting farther and farther apart. Locally, the matter self-gravitated and formed big blobs that we now call clusters. Then, within those blobs, the matter self-gravitated to form galaxies, and within those galaxies matter self-gravitated again to form molecular clouds and stars and planets and us. (For a poem depicting this history, see Appendix 4.)

Why is Lemaître's name not as well known as Hubble's? Part of the reason is that Lemaître doesn't have an important space telescope named after him. (I grew up in New York City and never heard of the Italian explorer Verrazano until the biggest bridge in the city got his name.) However, all astronomers and cosmologists know about Lemaître. And part of the reason we call it the Hubble expansion and not the Lemaître expansion is that Lemaître's analysis was based on an early version of Hubble's data that did not seem to support Lemaître's strong conclusions.

In the early sample of galaxies that Lemaître used, thirty-six of thirty-eight galaxies were receding (the Andromeda Galaxy is actually getting closer)—but their recessional velocity was not proportional to their distance as his model required; in fact, the points were almost consistent with randomness about the average. Lemaître seems to have believed that the disagreement was not a weakness of his theory, but just experimental error. The case would have to wait until better data were obtained. Moreover, perhaps to make it even more likely that nobody would notice his theory if it turned out to be wrong, he published it in an obscure journal.

In fact, the preliminary data could have been interpreted as refuting Lemaître's prediction. Had he prominently published words such as "I predict that when more accurate measurements are made, the galaxies will lie along a straight line, with their velocity of recession from us proportional to their distance from us," had he been so bold to make such a statement prominently, then perhaps we would now call it the Lemaître expansion. And Lemaître would have a telescope named after him.

In the Beginning . . .

In the Lemaître model, now accepted by most experts, a key feature is the expansion of space. The notion of expansion, of course, owes its birth to Einstein's rubbery-space ideas and, more immediately, to the equations of general relativity. But drifting into broader thought for a moment, it is fascinating to ponder the philosophical implications.

The Big Bang was not an expansion of matter within space, but the expansion of space itself. Space can be created, is being created, all the time as the universe expands. What happened at the "moment" of the Big Bang? Did space even exist before then?

My favorite answer to that question (based on speculation, not on scientific knowledge) is no, space did not exist prior to the first moment of the Big Bang. Where did it come from? Obviously, that question cannot be answered, because any answer assumes that there was a there there (using Gertrude Stein's phrase). But with no space, there is nothing for anything to come from. We could assume (to quote Rod Serling) that there is a "fifth dimension, beyond those known to man." Maybe that's where space originated, but that's really just a cop-out. So let's cop out in a different way, by ignoring the question and asking others.

Physicists tend to think of space not as emptiness, but as a substance. It is not a matter substance, but something more fundamental. It can vibrate, in many different ways. Vibrating space manifests itself as matter and energy. One mode of vibration manifests itself as a light wave; another mode of vibration is what we call an electron. If space did not exist before the Big Bang, then nothing would be able to vibrate, so neither matter nor energy would be possible. The creation of space is what made matter possible. Before space was created, none of the things we think about as "real" existed. We have no way to describe them.

I emphasize that these ideas are not part of science. They are merely the musings of a scientist. I'm sure I'm not the first scientist to have them. They are inappropriate for the scientific literature. But they are the sorts of ideas that scientists play with when they relax from the rigor of their profession. Maybe they will lead somewhere, but for now they are just reveries.

Space and time are linked by relativity. We don't live in space and time, but we live in space-time. Now think of the philosophical implications of that. If space began at the Big Bang, if space was created, maybe the same was true for time. Neither space nor time existed "prior" to the Big Bang; in fact, in this picture the word *prior* has no meaning. The question of what happened before time began is mean-

Figure 12.4. Calvin and Hobbes drop through a wormhole.

ingless, because there was no before. It is like asking, What happens if you put two objects closer together than zero distance? What happens if you cool a classical object below absolute zero, so that its motion is slower than no motion at all? These questions can't be answered, because they make no sense.

Augustine would have been comfortable with these thoughts. He argued that God's existence transcended time, that God existed outside of time. I suspect that if Augustine were alive today, he would preach that it was God who created both space and time.

Calvin—not the religious leader, but the cartoon character—wrestles with the meaning of time as he falls into what appears to be a wormhole in Figure 12.4. Hobbes seems to care more about the *now*.

The Solution to the Riddle

With the discovery of the Hubble expansion, we had an explanation for why the universe is so orderly— the condition that Eddington invoked to explain the arrow of time. The early universe, whether you think of it as a compact chunk of rock floating about in an otherwise infinite space, or as a Lemaître model with mass filling the entire universe, was compact. As space was created around the matter, there was more room, and that means more possible ways to distribute the matter and energy.

The expansion of space meant that the matter was in a relatively low-entropy state, compared to what it could be. The creation of space meant that there was a lot of empty space for additional accessible states, for additional entropy. And the universe, only 14 billon years old, has not yet had a chance to occupy the most probable high-entropy state. This idea—that although entropy continues to increase, the maximum allowed value for the entropy of the universe increases even faster—may have been first articulated by David Lazer, a physicist at Harvard.

The following example illustrates how expansion can create room for more entropy. Take a cylinder of gas, filled on one end, with a piston separating it from the vacuum on the other side. Assume this cylinder has been sitting for a while, so the gas within has reached the state with maximum entropy. Suddenly, very suddenly, move the piston, giving double the room for the gas to occupy. Do it fast enough that for a moment the gas is still on one side of the cylinder and the vacuum on the other. Now the gas is no longer in a maximum entropy state. It will not stay at one end, but will flow, fill the empty part, and expand until the new higher value of entropy is reached, with the gas filling the entire newly enlarged cylinder.

In a sense, that is what happened in the Big Bang. More space was provided, and the matter, which might have been at maximum entropy in the old small space, was no longer at the maximum entropy possible for the new larger space. It wasn't the matter that changed, but the number of possible ways it could fill the universe. That explanation provides an answer to the mystery of the current low entropy and—according to Eddington—gives an unambiguous direction to the arrow of time. Of course, in science when you answer a question, you often raise many more questions. No longer do we need to ask why we are in a low-entropy state. Now we need to ask, Why did the expansion happen? What caused it? Will it ever stop?

Can we ever obtain the final answer? I think not. We keep discovering new things, and they affect the answer. One recent discovery, *dark energy* (something I'll talk about later), dramatically changed the equations for the future expansion of the universe. We understand the laws of physics pretty well, but the knowledge of the universe and what it comprises is still new and uncertain. Maybe in a few decades or a few hundred years we'll discover something else completely new about the way the universe expands, and that will change our conclusions yet again. I think we can take joy in the speculation that we are not running out of important things to discover.

According to an ancient Greek myth, Sisyphus, the king of Corinth, was condemned forever to roll a huge stone up a hill in Hades, only to have it slip away and roll down again before it reached the top. His effort will never end. The great existential philosopher Albert Camus saw a parallel to life: we are born, we live, we die—to what end? Camus declared that the living of life was the end, and he concluded that Sisyphus is happy.

One might say the same for scientists. We can never answer all the questions; answer one, and a few new, harder ones pop up. Another classical analogy would be the heads of the Hydra; for every one you cut off, two grow back. Scientists love that. We'll never be out of work to do. It's what makes us happy.

13

The Universe Erupts

The physics of creation—the nature of the Big Bang . . .

A mighty flame follows a little spark.
— *Dante Alighieri*

This is the signal primeval.
The murmuring microwave background,
Cast by ylem of eld at three degrees Kel
Indistinct in the starlight.
— *Apologies to Henry Wadsworth*
Longfellow

A marvelous result of the Lemaître model is that it gives us a way of looking back in time—way back, way way back. I've personally looked back 14 billion years.

You look back in time all the time. When you look at people standing 5 feet away, you are not seeing them the way they are, but they way they were five-billionths of a second ago; that's how long it takes light to travel 5 feet. When you look at the moon, you are not seeing it the way it is now, but the way it was 1.3 seconds ago. When you look at the sun, you see the way it was 8.3 minutes ago. If the sun exploded 7 minutes ago, we wouldn't know, nary a hint, not yet.

The most distant and ancient signals we have observed so far are the cosmic microwaves, the signal primeval. We believe that they began their journey 14 billion years ago, and when we look at them (using a microwave camera), we see the way the universe looked back then. Of course, that light (microwaves are low-frequency light) is showing us what existed both a long time ago and very far away; it traveled 14 billion light-years of distance to reach us.

To say we are looking back in time, we have to make the assumption that the distant universe, 14 billion years ago, was very similar to what our nearby part of the universe was like back then. As I said earlier, that postulate is given a fancy name: the *cosmological principle*. Specifically, it asserts that the universe is homogeneous (like homogenized milk; it is uniform in composition without big lumps) and isotropic (no special directions; no large-scale organized motion; it isn't, for example, spinning). If you don't want other people to realize that you are making a drastic assumption, call it a principle. *Cosmological principle* is an awesome name; if you called it the raisin bread model, it wouldn't be so compelling. *Perfect cosmological principle* is an even more awesome name for an extension of the "ordinary" cosmological principle, but it turned out to be wrong; I'll discuss that in a moment.

There is good evidence that the cosmological principle is roughly true, at least true enough for our purposes. As we look around the universe, particularly nearby, we find things that are very similar to those in our neighborhood. We are in the Milky Way Galaxy (all the individual stars you can see with the naked eye are in this multiple-hundred-billion-star spinning clump), but there seem to be a huge number of similar galaxies out there, spread out and spreading further throughout space. Pick a small region of sky and, using our best telescopes, count the galaxies and extrapolate to the regions unexamined, and you conclude that there are over a hundred billion visible galaxies—most with fewer stars than our Milky Way has.

The image in Figure 13.1 was taken by the Hubble Space Telescope, at its maximum magnification. There appear to be about two thousand stars in this picture, but virtually all of those little spots are actually galaxies containing typically billions of stars each. The image was taken by peeking between the stars of our own Milky Way Galaxy, peering out beyond. At least two nearby stars are in the photo; you can spot them from the small cross-diffraction patterns that their point-like size gener-

Figure 13.1. The most distant region of the universe ever photographed. Virtually all the bright spots are galaxies, not individual stars.

ates. The most distant galaxy in the image is about 12 billion light-years away, meaning that we are seeing it the way it was 12 billion years ago. If the cosmological principle is accurate, then studying such galaxies can tell us what the Milky Way was like back then.

Although there are clusters of galaxies in space, those clusters seem to be everywhere, with an approximately uniform density. Microwave measurements made by my team at Berkeley in the 1970s showed that on the very large scale, the universe is uniform to better than 0.1 percent. Recent measurements made by the WMAP (Wilkinson Microwave Anisotropy Probe) satellite show that the universe is uniform to better than 0.01 percent, although at higher sensitivity, we do see nonuniformity.

The Big Bang Fireball

The most compelling evidence, the discovery of which confirmed the Big Bang, was the microwave remnant of the explosion. Had this signal not been found, the Big Bang would have been falsified; it would have been shown to be an incorrect theory. Princeton physicists Robert Dicke and James Peebles worked out the theory in the early 1960s; if the Big Bang hypothesis was right, these microwaves should be observable. If they found them, the discovery would be one of the greatest of the twentieth century, comparable in wonderment to Hubble's discovery of the expansion. They put together a team that included Dave Wilkinson and Peter Roll, and set out to build an apparatus capable of finding the evidence.

The theory was straightforward—well, as straightforward as cosmological theory based on general relativity gets. It was an elaboration of the original Big Bang theory formulated by George Gamow and Ralph Alpher. In the early universe, when space was 30 trillion times more compressed than it is now, the matter that filled it—the same matter we see today in stars and galaxies—was enormously dense and hot. The entire universe was filled with plasma as fiery as the present surface of the sun and, like the sun, also filled with intense light. Gamow and Alpher called this hot plasma *ylem*.

Gamow claimed that *ylem* was a Yiddish word for "soup," but I can't find it in a Yiddish dictionary; maybe it was a variant dialect. Alpher wrote that it was an obsolete word found in *Webster's New International Dictionary* referring to the "primordial substance from which the elements were formed." However, I could not find the word in either the 1913 or the 1828 edition of *Webster's Revised Unabridged Dictionary*. The *Oxford English Dictionary* gives one reference, to John Gower's 1390 *Confession amantis*, III.91, with the Middle English quote "That matere universall, Which hight Ylem in speciall."

They may have coined a new meaning for the word *ylem*, but neither Gamow nor Alpher made up the name *Big Bang*. That was Fred Hoyle, an eminent astronomer who didn't believe the theory and called it that to make fun of it. Perhaps to Hoyle's chagrin, Gamow readily accepted the name and used it himself. As another example of Gamow's sense of humor, when he and Alpher wrote their paper, Gamow added the name of his friend, the distinguished physicist Hans Bethe, as an author, even though Bethe had made no contribution, didn't give permission, and didn't even know he was a coauthor until the paper was published. Gamow then explained that it was a joke; he could not miss the opportunity to have a paper authored by Alpher, Bethe, and Gamow, since that combination was so reminiscent of the first three letters of the Greek alphabet: alpha, beta, and gamma. The work is sometimes still referred to by those letters, as the "$\alpha\beta\gamma$ paper."

Gamow was also a popularizer of science. Looking back as I write, I realize now that the book I loved so much as a teenager, Gamow's *One Two Three . . . Infinity*, was part of my inspiration for *Now*. I had also read Fred Hoyle's 1955 book *Frontiers of Astronomy*, which includes his defense of the "steady state" theory, his proposed alternative to the Big Bang. Hoyle argued that the expansion of the universe was an illusion, that matter was being continually created and destroyed, and that the universe wasn't changing. (As a kid, I had no opinion about who was right.)

Hoyle devised what he called the *perfect cosmological principle*, the claim that the universe was not only homogeneous in space, but also didn't change with time. In retrospect, I find it particularly interesting that Hoyle invoked *Occam's razor*, the scientific principle that the simplest theory is the right one, to claim that his theory was better than the Big Bang. One key lesson to take from this history: beware of *principles*. They are assumptions, not always based on fact. Another lesson is that *Occam's razor* is often a poor guide to truth.

When Alpher and Gamow first proposed the Big Bang theory, there was no way to test it, to falsify it—but Dicke and his team had found a solution to that challenge. According to their calculations, a half-million years after the Big Bang came a key moment, when the expanding space had cooled to the point that the plasma became transparent. At that moment the intense sun-like light suddenly could move freely, no longer bouncing back and forth off electrons, and it has moved without deflection ever since. That light from the early fireball is what the Princeton researchers wanted to find. They expected to be able to see it coming from all directions, because the Big Bang was completely uniform; that's the cosmological principle. The light would have traveled about 14 billion light-years, taking 14 billion years to reach us.

Of course, at our location matter was also hot and bright and full of light 14 billion years ago, and that light has been flying outward ever since. Right about now, light emitted from our location is reaching the very distant matter whose light is currently impinging upon us.

Because of the rapid expansion of the universe, the bright radiation emitted 14 billion light-years away has undergone a color shift; its source, that distant hot matter, was receding rapidly away from our current location (because of the Hubble expansion), and the light underwent a Doppler shift (in much the same way that Doppler radar detects your velocity by seeing a frequency shift). As a result, in our proper frame of reference the radiation should be not at the frequency of visible light, but at the frequency of microwaves, similar to those generated in your microwave oven but much less intense.

As Dicke, Peebles, Roll, and Wilkinson were preparing their instrument to search for this primeval signal, the team of Arno Penzias and Robert Wilson of Bell Telephone Laboratories was pointing a big and sensitive microwave antenna toward space. They were not looking for the Big Bang, but rather expected that by aiming at the blank sky they would be assured of having *no* signal; that way, anything that came down their receiver would indicate the intrinsic electronic noise generated within their apparatus. Their goal was to reduce this noise as much as possible.

They reached a limit, a certain amount of noise, amounting to about 3 degrees Kelvin (they measured noise in terms of temperature increase), that they could not get rid of. No matter what direction they pointed their antenna, it still showed the 3-degree noise. They concluded that the noise must be a signal coming from space, but they had no idea what it was, where it came from, why space was radiating it, or what caused it.

Actually, it was absurd to have a signal coming from space, uniform from all directions. At least, it was absurd back then. Penzias and Wilson must have had an extraordinary understanding of their instrument to reach such a patently ridiculous conclusion. Virtually any other experimenters would have decided that radiation that didn't depend on direction must have been coming from within the apparatus.

While his team was preparing its own instrument, Peebles gave a talk on the team's predictions. One of those who heard the talk was Ken Turner, who told Bernard Burke about it, who told Arno Penzias. Penzias called Dicke. Dicke's team happened to be in the room when Penzias called. "We've been scooped," Dicke told the team.

When Penzias and Wilson published their paper announcing the discovery, they made no mention of the Big Bang. Their paper was titled, diffidently, "A Measurement of Excess Antenna Temperature at 4080 Mc/s [megacycles per second]." They simply stated in their paper, "A possible explanation for the observed excess noise temperature is the one given by Dicke, Peebles, Roll, and Wilkinson (1965) in a companion letter in this issue." But within a year, the microwave radiation was considered the definitive evidence for the explosive origin of the universe. It was a prediction come true. The Big Bang had been observed.

Because of that talk given by Peebles, and the scientific grapevine that relayed the information to Penzias, the discovery was made by Penzias and Wilson, not by the Princeton team—which confirmed the signal a few months later. Penzias and Wilson shared the Nobel Prize for their work; the Princeton team got no such official recognition, except from peers (such as me). The award should have gone at least to Penzias, Wilson, Dicke, and Peebles, but Alfred Nobel's will prohibits it from being shared by more than three people.

The Quest for the Beginning of Time

In 1972, soon after I had earned my PhD at UC Berkeley working in elementary-particle physics, I was about to enter the same field and embark on my first major independent scientific project, based on my own work and interests and abilities and hopes—my first project done independently of my mentor Luis Alvarez. I had read a book by Peebles, *Physical Cosmology*, and I decided to try to observe the microwave radiation from the Big Bang. I wanted to see what the universe looked like 14 billion years ago and to test the validity of the cosmological principle.

The project I started eventually led to a map of the early universe, showing what it looked like when it was still a baby, only 0.00004 as old as it is now. For comparison, if you are twenty years old, then 0.00004 of your life corresponds to your first six hours.

Penzias and Wilson had determined that the microwaves were uniform to about 10 percent accuracy. They had detected no *anisotropy*— that is, no difference in intensity at different directions. Additional experiments had pushed that limit down to about 1 percent. At 0.1 percent there should be an anisotropy, from the motion of the Earth through the cosmos. Just as more water hits your face than the back of your head when you run through rain, the intensity of the microwaves should be slightly greater in the forward direction. The intensity should show a clear cosine dependence on direction, the same shape that you get from running through rain. And beyond that, at the 0.01 percent level, we might see remnants of the early clumping that led to clusters of galaxies.

In his book, Peebles called the motion of the Earth with respect to the distant universe "the new aether drift." It was not a measurement with respect to absolute space; Einstein had shown that that was impossible. But there is only one reference frame in which the matter of the universe is totally symmetric, uniform around you—the one reference frame in which the cosmological principle holds. That is the "canonical frame" of the Big Bang theory, Lemaître's frame—the one in which every galaxy is almost at rest, and in which the universe is expanding, not through motion of galaxies, but by the expansion of space between the galaxies.

To make the measurement, I concluded we would have to observe two frequencies simultaneously—one to quantify microwaves emitted from our own atmosphere, the other to detect the cosmic signal. The experiment would have to be done at high altitude, possibly on a mountaintop, but more likely in a balloon or airplane. Having done experiments with balloons, I decided that they were too difficult (for example, prone to crashes). I also believed that to keep things simple, the project would have to be done using ordinary room-temperature devices, not cooled

Figure 13.2. Marc Gorenstein (left) and the author installing the microwave detectors on the back of NASA's U-2 aircraft, 1976.

low-noise detectors. Using relatively warm antennas would require that the receiving radiometer have excellent thermal conduction properties, so that the instrument itself would not introduce an apparent anisotropy. So, for the first time in my life, I studied the design of heat flow.*

I asked another physicist at Berkeley's Space Sciences Laboratory, George Smoot, whether he would like to join the team, and he agreed. Hans Mark, director of the NASA Ames Research Center, offered the use of NASA's U-2 research aircraft, and we adapted our instrument to fit in it. To eliminate radiation from the Earth, I concluded that we needed to use microwave horns that were *apodized*—that is, horns that used an advanced optics concept to reduce signals coming in at wide

---

* For a longer list of the problems we had to address, see my May 1978 *Scientific American* article: "The Cosmic Background Radiation and the New Aether Drift."

angles. Smoot found a published apodized design that should work; we built several horns, and I tested them in a microwave lab on the Berkeley campus, helped by my first graduate student, Marc Gorenstein, who eventually earned his PhD on the project.

It was a long, hard slog, but after a series of U-2 flights, we found that the radiation was not completely uniform. The brightest region was just south of the constellation Leo, and the darkest was exactly opposite, in the constellation Pisces. Between those limits the variation was smooth, proportional to the cosine of the angle from the Leo maximum—a clear confirmation that the radiation was due to the motion of the Earth with respect to the distant matter of the universe. From the amplitude of this *cosmic cosine*, I calculated the velocity of the Milky Way Galaxy and found it to be close to 1 million miles per hour, a nice round yet impressive result. The cosmic cosine is shown in Figure 13.3, with stars and constellations superimposed. This figure

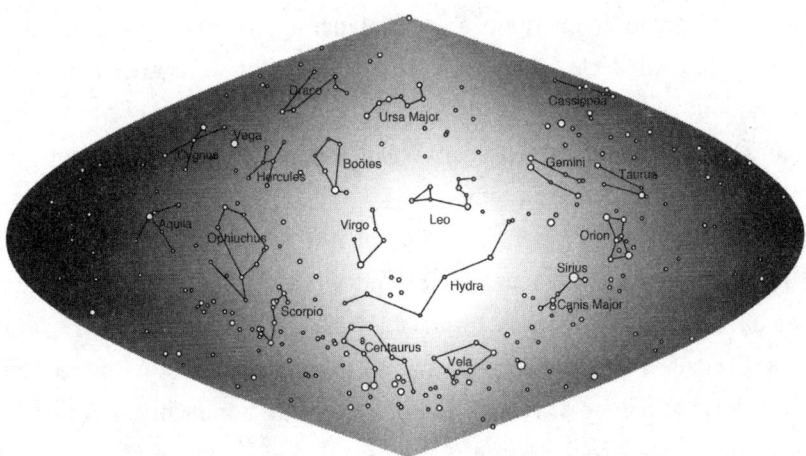

Figure 13.3. The cosmic cosine. A star map with the intensity of the cosine variation of the microwave radiation added as background brightness. This is what the sky would look like if you could see microwaves. The strongest signal (just south of the constellation Leo) indicates the direction of our peculiar motion. At other locations on the map, the intensity is proportional to the cosine of the angle from the maximum.

is based on updated data from more precise measurements made with the recent WMAP satellite, but they agree within the uncertainties of the values we published in 1976.

If we are moving a million miles per hour, how could our galaxy be at rest in the Lemaître picture? The answer is that it isn't. Lemaître's model allowed for individual galaxies to have a small local movement, called *peculiar motion*, perhaps swirling around a local cluster, or—as in our case—the Milky Way being attracted by the gravity of the nearby Andromeda Galaxy. Lemaître simply assumed that such local motion was small, on average, and in random directions.

We Are a Quantum Fluctuation

The next step was to put the project into a satellite, to completely eliminate the confusing microwave signal emitted by the atmosphere. At that point, Smoot took over leadership (and fund-raising) of the project, and I gradually dropped out. It was clear to me that he didn't need my help, and I wanted to avoid the NASA bureaucracy. Indeed, that proved to be the main obstacle; changes in the instrument were minor, but the officialdom of the US government was major. George went ahead, but it took more than fourteen years (1978–92) to get the instrument into space and make the measurements.

There was no fundamental reason for it to take that long—four years would have been more reasonable—but working with the government was bureaucratically complex and often *not* driven by the needs of science. Every few years, it seemed, NASA told Smoot that the instrument had to be modified to fit yet a different spacecraft. For a while, it was to go on an unmanned rocket; then NASA decided that more science needed to be put on the space shuttle (to help justify the shuttle's expense!), and that delayed the project. In addition, flying with humans meant that the instrument had to pass many additional tests, to make sure it would not endanger the lives of astronauts. Then NASA changed its mind again and decided to put it back on an unmanned rocket, but it would have to fit together with a totally different experiment designed to measure the spectrum (intensity at different frequencies) of the radiation.

Out in space, with no interference from the atmosphere, the results that Smoot and his new team obtained were thirty times more sensitive than our U-2 results, and for the first time they observed the intrinsic anisotropy, the anisotropy that showed a slight departure from the cosmological principle. The lumps and bumps they saw were just as expected from the Big Bang. That theory assumed a universe that started out pretty uniform but not perfectly so; "quantum fluctuations" from Heisenberg's uncertainty principle would cause small lumps that would grow from local gravity, forming structures that would eventually become large-scale clusters of galaxies.

It was fascinating—stunning—that cosmology, the realm of the super big, could best be understood through the behavior of quantum physics, a realm that had previously been found only in the very small. Stephen Hawking called the discovery "the most exciting development in physics" he had experienced in his life. The cosmic microwave radiation, whose discovery had confirmed the Big Bang theory, was now the most detailed source of information about the nature of the first half-million years of the explosion. Smoot was awarded a Nobel Prize for his work.

We hadn't quite reached the beginning of time—only a half-million years after the beginning. On a human scale, a half-million years sounds like a lot, but compared to the 14,000 million years that have come since, we had managed to photograph a universe that was like a newborn baby a few hours old. And most important, this was not a theory; it was an actual observation.

The results were later improved by a measurement made by the WMAP satellite (which was named for one of the original members of the Princeton team). The pattern observed, the detailed structure of the view of the universe a half-million years after the Big Bang, is shown in Figure 13.4. Think of this as a photograph of the most distant sky, in microwave light invisible to the eye, showing the structure of the universe as it was 14 billion years ago at a distance of 14 billion light-years.

Figure 13.4. A photograph of the early universe, taken using the WMAP microwave camera. The smooth cosmic cosine was removed from the image, since otherwise it would completely dominate it.

Looking at it, you can see that even after a half-million years, the universe was no longer completely uniform but had begun to clump up.

The Thurston Universe

I had the fortune of having as a brother-in-law one of the greatest mathematicians of recent times, Bill Thurston. Bill and I lived a few blocks from each other in Berkeley, and we had many conversations about careers (as a graduate student, he was convinced that he would never find a good job), mathematics, and physics. He was fascinated with my description of what we knew about the universe. He asked me if anybody had seriously considered a multiconnected universe? Did he mean wormholes, passageways that potentially connected one part of the universe with another? No, he had something far simpler and much more elegant in mind.

Bill was ultimately most famous for his advances in topology, complex geometries that went far beyond our normal imagination. He told me that he had actually mastered the skill of being able to think in four dimensions. Few people believed him, until he produced a vast array of wonderful theorems that he claimed he had discovered by simply looking at surfaces in 4D space in his mind. It turns out, oddly, that math problems in three and fewer dimensions are relatively easy, and problems in five or more dimensions are also relatively easy, but dealing with four dimensions is very tough. Thanks to his work in four dimensions, Bill was to win the Fields Medal, the "Nobel Prize of mathematics," before he turned forty.

In topology, you can move forward in space and then discover you are back where you started. Such an outcome is trivial for a curved space (such as the surface of the Earth), but it can also work for uncurved space. Uncurved space is usually referred to as "flat space" by cosmologists, even though it exists in three dimensions. All it means is that, on the large scale, light does really travel in straight lines, not curved ones; ordinary geometry still works; the angles of a triangle still add up to 180 degrees.

Bill's question was whether the actual universe was a simple or multiply connected universe. He wanted to know whether any measurements in cosmology could rule out the multiply connected version. I could not think of any. Could any confirm it? That was worth some serious thought.

I consider this "Thurston universe" (as I call it) to be a great and wonderful out-of-the-box speculation. It is multiply connected, like wormholes, but with no strong spatial distortions, and it has the exciting advantage of being testable. I certainly don't consider it nearly as crazy as the eleven-dimensional space-time used in some string theories.

I spent several weeks trying to prove that the Thurston universe wasn't true, and looking for ways to discover it if it was. To verify it, I could look into distant space and see the Milky Way, our own galaxy. Maybe one of those galaxies in the Hubble deep-space photo (Figure 13.1, page 138) is actually us! I would be seeing it, however, not the way it looks now, but the way it looked a billion years ago. Wow! If Bill's speculation was true, we would be able to look back in time in a way that would require no assumption of a uniform universe. We could actually *see* ourselves. The challenge was *recognizing* ourselves. Galaxies evolve a lot in a billion years, and galactic clusters evolve too. I thought hard about testing his idea but eventually gave up. Of course, that was in the early 1980s; instruments change. I'm looking at it again now.

This example illustrates what experimenters do in their spare time. My physics mentor, Luis Alvarez, made a point of setting aside Friday afternoons to think crazy. Unless you set the time aside, you'll never find time to do it. It's just like exercise.

14

<hr>

The End of Time

*Now that we know what happened in the past 14 billion years,
what can be said about the coming 100 billion? . . .*

> To see a world in a grain of sand
> And heaven in a wild flower
> Hold infinity in the palm of your hand
> And eternity in an hour.
> — *William Blake,* Auguries
> of Innocence

Back in the late 1990s, when I was teaching cosmology, I told my class that although I couldn't tell them the long-term future of the universe, I was confident I could predict that a major discovery was imminent. Within five years, I said, we would know whether the universe is infinite or finite, whether it will continue to expand forever or eventually stop and crash back on itself in a Big Crunch. And if that crunch happens, it is reasonable to think that it could be the end of both space and time—forever, if use of that word makes any sense when time no longer exists.

I also said it was possible that we would wind up being delicately balanced on the dividing line between infinite and finite (both space and time), so although we would have a precise estimate of the universe, it would not really answer the question of whether forever is truly forever.

I was absolutely certain I was right in my prediction that we would soon know the answer. The reason was that I had founded the scientific experiment myself, the one that would determine the answer. And I had confidence in my former student, Saul Perlmutter, who had taken over leadership of the project.

The Quest for the End of Time

The microwave project I described in the previous chapter was looking at the structure of the Big Bang, the way the universe was structured in its very early moments. The goal of this next experimental project was to ascertain the future of the universe. The way to do that was to determine, much more precisely than ever before, the exact behavior of the Hubble expansion.

Theory predicted that the expansion would be slowing because of self-gravity, the mutual attraction of the galaxies that were rapidly getting farther and farther apart. We could measure the slowing by examining the Hubble expansion in both nearby galaxies and distant ones. The distant galaxies would show Hubble's law the way it was billions of years ago, and we should be able to see how much the expansion had slowed since then. We could measure the velocities of the galaxies using the standard of police radar: the Doppler shift.

The hard part was to obtain a good measure of the galaxies' distance. I had concluded that supernovas held the key to accomplishing that. Once we could detect the deceleration of the universe, we could calculate whether or not the expansion would go on forever. The computation was very similar to calculating escape velocity. Would the expanding galaxies escape, or would they fall back into a Big Crunch?

Cosmologists had assigned a symbol to the deceleration parameter: capital omega, after the last letter in the Greek alphabet, Ω. Our goal was to determine omega, and I tentatively named our experiment the "Omega Project." Omega would tell us about the possible end of time.

The Omega Project was inspired by a talk by Robert Wagoner that I had heard at Stanford in 1978. He pointed out that the intrinsic brightness of distant type II supernovas could be determined by observing the rate of expansion of the supernova shell and the time it took to expand; velocity multiplied by time would give its size. If we could find distant supernovas, deduce their brightness, and measure their velocity by the Doppler shift of the galaxies that held them, then we could use them as "calibrated candles." The observed brightness, compared to their intrinsic brightness, would determine their distance.

The key was to obtain data from a large number of distant supernovas. But supernovas are rare phenomena; in any one galaxy, you'll find one such explosion only once every hundred years or so, and you have to catch them in the first few days if you want to use them in any meaningful way. You have to watch thousands of galaxies, coming back to look at them every few nights, if you want to see them during the critical expansion phase.

When I told my mentor and former thesis adviser Luis Alvarez about Wagoner's talk, he said that Stirling Colgate, a physics professor at New Mexico Tech, had recently set up a project for the automated discovery of supernovas. I visited Colgate and learned that he had abandoned the project as too difficult. But he encouraged me to try, and he gave me lots of advice about how to succeed where he had failed.

I needed a telescope and a very powerful computer to run it. Fortunately, my discovery of the cosine anisotropy of the microwave radiation had been rewarded with the Alan T. Waterman Award of the National Science Foundation—$150,000 of unrestricted research funds I could use on any project I devised. How wonderful that award was! I could start my supernova search without having to prove to referees that I was qualified to do so. The Waterman Award made the project possible. I used it to buy the computer I needed (in those days, powerful computers were still very expensive) and to hire a recently graduated physicist, Carl Pennypacker, to help.

The project was daunting, and I had to obtain additional support. We got that, only to lose it. Twice the project was canceled by administrators (once by the director of the physics division at the Lawrence Berkeley

Laboratory, and once by the director of the Berkeley Center for Particle Astrophysics). But I managed to raise funds and keep it going. It was great to have tenure; my continued job (and salary) did not depend on my following orders from any boss. It seemed to me that once again the bureaucratic challenges were greater than the physics ones, as had been the case for George Smoot with NASA.

In 1986, eight years after I began the supernova search, my fourth graduate student, Saul Perlmutter, joined the team as a postdoc (a hire right after he received his doctorate with me). He quickly demonstrated amazing leadership qualities. Saul completely rewrote our automated computer software. By looking at hundreds of galaxies, repeatedly, we began to find supernovas. By 1992, we had reported the discovery of twenty supernovas, including the most distant one found to that date.

Most of our supernovas were nearby, by cosmological standards. Saul and Carl wanted to jump ahead and begin the search for extremely distant supernovas—those that would require bigger telescopes to find but that offered true hope of seeing the expected deceleration of the expansion. Although I was dubious, I believed in them, and I approved the new direction. Saul devised an innovative way to transmit data over the slow international networks of the time, using the math of *fractals*. As far as I know, he was the first person to use this advanced method for a scientific measurement; it is now in wide use.

Then Saul solved a key problem, one that had totally frustrated me. He devised a scheme for discovering many supernovas on a single night just prior to the new (dark) moon, and a schedule that would allow follow-up at a large telescope (such as the space telescope) on a subsequent dark night. In my mind, that simple breakthrough approach was what made the project work.

It may seem surprising to the nonexperimentalist that I started the project without knowing how we could address the major issue of follow-up measurements. However, I had learned from Luis Alvarez that such daring was necessary—or you'd never tackle a great challenge. You have to have confidence that you (or a member of your team) will invent the solution

Figure 14.1. The author in 1984, blocking the view of the small telescope we initially used for the supernova project.

when needed. If I hadn't had the funds from the Waterman Award, I would not have been able to take such an adventurous approach; referees would have demanded answers to every question and rejected all requests for funds until we had found answers that satisfied them.

Saul presented the solution during one of our closed meetings at which outside referees evaluated our work to advise whether additional funding should be approved. This was the same group that had previously recommended that the supernova project be canceled. After Saul's presentation, it was obvious to the review committee that the project would succeed. In fact, Robert Kirschner, one of referees on the review team, found the idea so convincing that after he returned home from the review, he helped create an independent research group to race our Berkeley team to the answer.

In my mind, Saul had become the true project leader, so in 1992, fifteen years after the project began, and six years after he joined it, I asked Saul to take over management. I gradually drifted away to found other projects. Five years later, Saul had made so much progress and was getting so close to an answer that I told my class at Berkeley we would soon know whether time will last forever or will end in a Big Crunch.

The Accelerating Universe and Dark Energy

In 1999, Saul and his group, a team that had expanded into an international collaboration, made an astonishing, unbelievable discovery. Their measurements of Hubble's law, done with substantially greater precision than ever before, looking at very distant galaxies, detected a departure from the law. The universe was not slowing down as expected from mutual gravity. There was a greater force, something that was making the expansion speed up. It was a completely unexpected and startling development. When Saul showed it to me, I was highly skeptical. Saul and his team worked very hard to find anything that could lead them to a mistaken conclusion, but they failed. They could not falsify their own work. They had discovered that the expansion of the universe was accelerating!

Virtually simultaneously, the group that Kirschner had helped create announced similar results. A few years later, Saul shared a Nobel Prize with Brian Schmidt and Adam Riess from the rival team for the discovery.

With acceleration of the universe confirmed, the issue of the Big Crunch was settled. It will not happen. Space goes on forever, and so does time ... unless, of course, there is yet another phenomenon to uncover, something that perhaps hasn't yet shown its effect, but that could eventually turn the universe around. And eventually, the acceleration that Saul and his team discovered could give us an experimental test for a theory I'll describe later in this book, the 4D Big Bang theory, that gives my proposed explanation for the meaning of *now*.

I hadn't anticipated that the universe is accelerating. Nobody had. But I had correctly predicted to my class that we would know whether the universe will expand forever. And once Saul had released the results,

Figure 14.2. Saul Perlmutter (right) with the author, in the Swedish Royal Palace following the Nobel Prize ceremony in 2011.

before it got into the newspapers, I was able to tell my physics class that the answer had come.

Einstein's Greatest Blunder

The acceleration of the universe was readily accommodated by Einstein's general theory of relativity. Recall that before Hubble discovered that the universe was expanding, Einstein had assumed it was static, with galaxies simply staying where they were. To cancel their mutual gravitational attraction, he had introduced a *cosmological constant*, a repulsive force, to try to make the universe static (before the Hubble expansion was discovered). Hubble gave this constant the symbol Λ, the Greek capital letter lambda. It served as a repulsive force, a kind of antigravity, but it originated from empty space instead of from mass. I think of it as space repelling itself.

When Hubble discovered that the universe was expanding, the lambda term was no longer needed, and the cosmological community simply assumed it was zero. As I mentioned in Chapter 12, according to George Gamow, Einstein called the inclusion of lambda the biggest blunder of his life. Had he not added it, he could have predicted the *expansion* of the universe! In what I consider to be the greatest irony of Einstein's life, we now know that Einstein's bigger mistake was not in sticking lambda in, but in taking it out. Had he left it in, he could have predicted the *acceleration* of the universe. Einstein's greatest blunder was in calling the cosmological constant a blunder.

A convenient way to include the cosmological constant lambda in the equation of general relativity is to move it (mathematically) into the energy part of the equation, by combining it with the quantity T that represents the energy density. That's equivalent to thinking of lambda as an energy term. In fact, that approach has now become conventional, and the presence of the cosmological constant is described by saying that empty space is full of a *dark energy* with a density and pressure that depends on the value of lambda. When the cosmological constant is included in this way, the equation of general relativity is unchanged; the lambda term is not there, but the energy and pressure of empty space are no longer zero.

Dark energy, filling empty space, sounds like the aether all over again . . . and it is. Empty space in modern cosmology is hardly empty. In addition to dark energy, physicists now believe that "empty" space includes a *Higgs field* that serves the purpose of making particles appear to have greater mass than they would otherwise have. And Paul Dirac even proposed that empty space was filled with an infinite sea of negative-energy electrons—certainly the most astonishing proposal ever to come from an eminent physicist. (More about this in Chapter 20.) Vacuum is hardly empty.

One of the reasons theorists like to move lambda over into the energy term, making it dark energy, is that they already expected a similar term to appear, thanks to considerations of quantum physics. They expected that "quantum vacuum fluctuations" would carry dark energy and would come with negative pressure just like the dark energy discovered by Saul. So why don't we give them credit for having predicted dark energy? The reason is that they got the number wrong. Whereas we know that the dark energy that accelerates the expansion of the universe has a mass density of about 10^{-29} gram per cubic centimeter, the value predicted from the quantum physics theory is 10^{+91}. The theory is wrong by a factor of 10^{120}. This disagreement has been called "the worst theoretical prediction in the history of physics." The quantum theory prediction of the dark energy is wrong by a hundred quintillion googols.

Could quantum fluctuations still be the source of the dark energy? Maybe. Some theorists are looking to adjust their theory by the necessary amount, but nobody seems to have a plausible way of changing the numbers by such a huge factor. My own guess is that the correct value for the energy of the quantum fluctuations will turn out to be zero (once we have the correct quantum theory), and that the dark energy will turn out to be something completely different, something analogous to the Higgs field (which I will discuss in Chapter 15). That's just a guess.

Inflation

Rapid expansion of the universe, at speeds exceeding that of light, are a key part of the theory of *inflation*, an idea put forth by physicists Alan Guth and Andrei Linde and further developed by Andreas Albrecht, Paul Steinhardt, and others. The problem they addressed was the remarkable uniformity of the universe. If we look 14 billion light-years away, we observe the location that emitted microwaves 14 billion years ago. That radiation is just now reaching us, having traveled for 14 billion years. But if we look in the opposite direction, we see radiation coming from there too, also having traveled for 14 billion years.

Those two regions are 28 billion light-years apart. The universe is only 14 billion years old. So there was no time for a signal to travel from one side to the other. Even when they were close, in the early moments of the

Big Bang, they were moving apart too rapidly to be in contact with each other. So how did they "know" how to reach the same density, to have the same temperature and the same intensity of radiation? How could they be so similar when they didn't have time to contact each other, to reach equilibrium? Signals from places 28 billon light-years apart are observed to be extremely similar. How could they arrange that?

Guth and Linde showed it was possible that those distant points were once very close to each other at a time when the expansion of the universe was sufficiently slow that they could interact. They were close enough to reach a similar temperature and density, and then, suddenly, according to their theory, the nature of the vacuum changed and they separated at speeds enormously faster than lightspeed. Those points didn't move; their separation came from the rapid creation of space between them. This increase in space is what they called *inflation*. The math worked out. They had to postulate a new kind of field that was responsible, a field that changed as it expanded and eventually settled into a state in which inflation stopped, but such a field was readily accommodated into general relativity.

For many years, the idea of inflation was popular primarily because it offered the only known solution to the mystery of how the universe became so uniform. Inflation's answer: it was once all in intimate contact. Recently, however, some other predictions of inflation have been verified, including predictions of the kind of pattern seen in the microwaves. The idea of inflation became even more plausible when the present acceleration of the universe was discovered.

15

Throwing Entropy under the Bus

*I admit to my doubts about Eddington's explanation
of the arrow of time . . .*

The total disorder in the universe, as measured by the quantity that physicists call entropy, increases steadily as we go from past to future. On the other hand, the total order in the universe, as measured by the complexity and permanence of organized structures, also increases steadily as we go from past to future.

— *Freeman Dyson*

D o you feel, now, that the mystery of the arrow of time has been solved? Are you persuaded by Eddington's argument, by my attempt to reproduce it? Or do you feel, like the physics friends I've queried, that you're not completely sure it's right?

I have a confession to make. I think the entropy explanation of the arrow of time is deeply flawed, and almost certainly wrong. Writing several of the last few chapters—starting with Chapter 11, "Time Explained"—was difficult for me, but I wanted to give Eddington's best case before presenting my objections.

Are there alternative explanations for the arrow of time? Yes, several, including the possibility that quantum physics, a subject far more mysterious than relativity, sets the arrow. Another is that the arrow is determined by the creation of new time by the same Big

Bang that is constantly creating new space. I can't prove that either of those is right, but I am convinced that Eddington's explanation is wrong.

What tests do we apply to determine whether a theory is valid?

Successful Tests of a Theory

Look to Einstein for the standard of quality in theory. When he devised his initial theory of relativity, later called special relativity, he made definite predictions about the behavior of time and length. Ten years later he made additional predictions about the way these would vary in gravitational fields. In 1919, Eddington verified Einstein's predictions of the deflection of light from the sun. The first detection of energy-mass equivalence may have been a paper in 1930 by George Gamow pointing out that the "mass defect" seen in nuclei was related to the negative energy of the nuclear force. From Einstein's theory, Dirac predicted the existence of antimatter, which was discovered by Carl Anderson in 1932. In 1938, Herbert Ives and George Stilwell detected and verified Einstein's equations for time dilation. The mass-energy equivalence was dramatically observed in electron-positron annihilation in the 1940s. And all the standard relativity effects—time dilation, length contraction, mass-energy equivalence—are now seen daily in modern physics laboratories.

Einstein was very specific about the falsifiability of his theories. In 1945 there was a serious discrepancy between the age of the Earth (measured in radioactive rocks) and the age of the universe (determined by the Hubble expansion). When he updated his book *The Meaning of Relativity* that year, Einstein wrote,

> The age of the universe, in the sense used here, must certainly exceed that of the firm crust of the earth as found from the radioactive minerals. Since determination of age by these minerals is reliable in every respect, the cosmologic theory here presented

would be disproved if it were found to contradict any such results. In this case I see no reasonable solution.

Einstein did not have to retract general relativity; it was the experiment that was wrong, not his theory. Hubble had not recognized that he had confused two types of very similar stars when making his measurements. After this error was uncovered and revised calculations were made, the corrected age of the universe turned out to be greater than the age of the Earth, as it obviously had to be. But it is refreshing to read Albert Einstein stating that, if the experimental numbers didn't change, the theory would be proved wrong, with "no reasonable solution."

In the next paragraph, I'll list the predictions made by Eddington's 1928 theory for the arrow of time, including all the predictions later made by other theorists who worked on the theory.

[This paragraph intentionally left blank.]

That blank paragraph represents the predictions of Eddington and the other physicists who link the arrow of time to entropy. There were none; there are none. Modern authors who present the entropy theory of time's arrow sometimes admit to this deficiency. Sometimes they express optimism that predictions are just around the corner. But, as of the date of publication of this book, 2016, it has been eighty-eight years since the Eddington theory was proposed as the explanation for the arrow of time, and there has not been one experimental test—not accomplished, not even proposed.

Or has there? Certain effects, if they had turned out to agree with Eddington's entropy arrow theory, would have been widely cited as having proved it to be true. Yet when these effects are not seen, that negative result is not taken to be evidence against the theory. That's because Eddington's theory makes no predictions; it only "explains" phenomena. A theory that makes no predictions cannot be falsified. I suggest we use the term *pseudotheory* for claimed theories that can be verified but not falsified.

If time is related to entropy, would you expect to see some effects? Relativity is full of such phenomena. Local gravity affects the rate of clocks; should not local entropy do the same? When entropy of the Earth's surface decreases at night, shouldn't we expect to see a change in the rate of time, perhaps a local slowing? But it doesn't happen. Why not? If such slowing had been observed, it certainly would have been considered a triumph of Eddington's theory, even though he never predicted it.

According to the standard model, the increase in entropy of the *universe* is what determines time's arrow. So let's look at the entropy of the universe. Where is it?

The Entropy of the Universe

The entropy that was known to Eddington was the entropy of the Earth, the sun, the solar system, other stars, nebulas, starlight, and other things we can see and detect. Since his time, we have discovered that such entropy is a minuscule part of the total entropy of the universe.

The first revelation of enormous unanticipated entropy came when Penzias and Wilson discovered the cosmic microwave radiation. It doesn't have much entropy per cubic meter, but it fills all of space, unlike ordinary matter. As a result, we estimate that the entropy of these microwaves is about 10 million times greater than the entropy of all of the stars and planets combined.

How is this enormous entropy of the cosmic microwaves changing with time? Remarkably, it isn't. As the universe expands, the microwaves fill up more space, but they lose energy; the net result is that their entropy remains constant. This huge reservoir of entropy, so much greater than the entropy of the stars, is not changing. Yet time goes forward. Should this lack of change in entropy be taken as falsifying the entropy arrow?

Physicists believe that the universe has three other big tanks of entropy, but they have never been observed or verified; they are theoretical in nature. The first consists of the neutrinos left over from the Big Bang. They are almost as abundant as the microwave photons, but they interact with matter even less. There are three different kinds of neutrinos, and since they don't interact, their entropy is also constant and comparable to that of the microwave photons.

The second big source of hidden entropy resides in supermassive black holes. The entropy of a black hole was first calculated by Jacob Bekenstein and Stephen Hawking. Most theorists seem to accept their work, but it has not yet had any experimental confirmation. Because their work is at the very edge of our knowledge of relativity and quantum physics, it may prove important, whether it turns out to be right or wrong.

For the sake of argument, let's assume that Bekenstein and Hawking are right about this entropy. Such entropy is expected to increase as matter is accreted into supermassive black holes. And estimates for the present entropy of supermassive black holes suggest that it is billions of times larger than the entropy of the microwaves. The nearest supermassive black hole is buried in the heart of the Milky Way Galaxy and appears to be accreting matter. That means its entropy is increasing.

Assuming the Bekenstein-Hawking formula, the entropy of supermassive black holes totally overwhelms the matter, microwave, and neutrino entropy of the universe. Is it the black hole in the center of the Milky Way that determines the arrow of time on Earth?

Here is a key fact about that entropy. Nominally, that black hole is 14,000 light-years away. But the entropy is buried deep down near the surface of the black hole. Assuming the black hole has actually finished forming, that entropy is an infinite distance from us. In reality, the distance will just be very large, given approximately by the speed of light multiplied by the time since the black hole began to form. So, that entropy is at least billions of light-years away. How could such distant entropy have an effect on our time?

There may be one other, even larger store of entropy. It's in what is called the *event horizon* of the universe, about 14 billion light-years away. That entropy is increasing rapidly as the universe expands. But it is moving away from us at virtually the speed of light. And it is very far away.

Remember, there is no established connection between entropy rise and time flow; it is merely a speculation based on a correlation—the fact that both are moving forward. There is no actual *theory*, in the sense that general relativity is a theory. Maybe someday there will be a true theory. I can't rule it out, but it is hard to believe that such a theory will show that these remote entropies determine the arrow of time, or will link us to the nonchanging (and almost noninteracting) microwave entropy.

We all know that correlation does not imply causation. There is even a Latin expression for the error in thinking: *cum hoc ergo propter hoc.* Literally, it means "with this, therefore because of this." It refers to the mistaken assumption that two events that are correlated are necessarily causally related; that is, one caused the other. Use such logic and you might conclude that sleeping with shoes on causes hangovers, or that ice cream sales cause increased drowning, or something else equally absurd. Yet, too often, physicists don't recognize that they are falling into the same trap when they argue that the arrow of time is determined by entropy.

The great philosopher of science Karl Popper argued that for any theory to be considered scientific, it must be able to specify how it can be falsified. The entropy theory of time's arrow fails Popper's criterion.

Theories that are not falsifiable include spiritualism, intelligent design, astrology, and the linkage between the arrow of time and entropy. You can probably think of others. Of these, astrology comes closest to being falsifiable. A careful experiment done by Shawn Carlson (with me as his scientific adviser, and using part of my Waterman Award funds to pay for the astrological charts) was published in the prestigious journal *Nature*.* Shawn tested the fundamental thesis of astrology—that precise time of birth is correlated with personality traits—and he did it in a double-blind fashion that was endorsed (until his results came out) by some of the most respected astrologers in the world. (Yes, there actually are such people; most of them have PhDs in psychology.) After

*   Shawn Carlson, "A Double-Blind Test of Astrology," *Nature* 318 (December 5, 1985): 419–25; doi:10.1038/318419a0.

his results falsified that fundamental thesis, the astrologers expressed shock and disappointment (they really do take their field seriously), but none of them withdrew their acceptance of astrology. So astrology is falsifiable to scientists—and it has been falsified—but astrologers persist in their discredited discipline.

In a Greek legend, Antaeus was a wrestler who kept his enormous strength only as long as he was in contact with the ground. I imagine he was a metaphor for the gentleman farmer; once he no longer gets his hands dirty every day, the crops will fail. Antaeus's pastime was challenging all those who passed by him to a wrestling match. He always won, killed them, and then kept their skulls to build a temple. Antaeus finally fought Hercules. Hercules was losing, until he remembered the secret need for ground contact. He managed to lift Antaeus off the ground and then crushed him in his arms.

As an experimentalist, I sometimes sense the Antaeus effect. If I have not worked in the machine shop for several months, I forget why it takes ten minutes to put in a screw, and I can be overly harsh with my students. (A proper screw requires careful measurement, two drill holes, and finding the right screwdriver. It may take ten minutes to put in a single screw, but it takes only twelve minutes to put in five screws.)

Theoretical physics must keep in touch with the ground, by insisting on having testable and falsifiable experimental results. Had Eddington observed a different value for the deflection of light by the sun during an eclipse, Einstein would have been shown to be wrong. If high-speed particles did not have extended lifetimes, Einstein would have been shown to be wrong. If GPS didn't need to correct for both gravity and velocity time dilation, Einstein would have been shown to be wrong.

Indeed, Einstein's theory of Brownian motion appeared to have been proved wrong soon after he published it. A series of experiments falsified his work. It was during this period that Ludwig Boltzmann, the father of the then still disputed statistical physics, committed suicide. However, subsequent experimental work showed that there were flaws in the first experiments, and Einstein's predictions were verified. It took four years.

The God Particle Breaks Entropy's Arrow

Let me formulate another prediction, not made by Eddington but one that I think should follow from his theory. According to our standard model of cosmology, in the very early universe no particles had mass; electrons, quarks, and all others were as massless as the photon. This remarkable state of affairs provided a key to making the early universe work, to helping the grand theories of unification make mathematical sense. Later, as the universe evolved, the particles (according to the standard theory) "acquired mass"—through the so-called *Higgs mechanism*.

Simply, the Higgs mechanism means that the entire universe was suddenly filled with a Higgs field through a process called *spontaneous symmetry breaking*. The previously massless particles, when moving through this fluid, behaved as if they had mass. In the Higgs mechanism, mass is an illusion, although it retains all of the properties expected from relativity.

The theory predicted that a chunk of a Higgs field could be created in sufficiently energetic collision, and that prediction was confirmed on July 4, 2012, when CERN, the large particle research center at Geneva, announced the discovery. The image in Figure 15.1 shows explosive debris coming from the radioactive decay of a Higgs particle.

Leon Lederman, Nobel laureate for his discovery of the muon neutrino, and my teacher at Columbia, wrote a book about the Higgs titled *The God Particle*. He claims that the title was the idea of his editor; it probably did increase book sales by tenfold or more. The reason for this name is that the Higgs field gave particles their mass, and without that, there never could be atoms or molecules or planets or stars. That is true, although by that reasoning we could call the electron the God particle, since without electrons we likewise could not exist—or the photon, or virtually any other particle in our list of elementary particles. There is a consensus among physicists that the "God particle" is the worst possible thing to call any particle—even worse than naming two quarks after "truth" and "beauty" (which some physicists tried to do). Nevertheless,

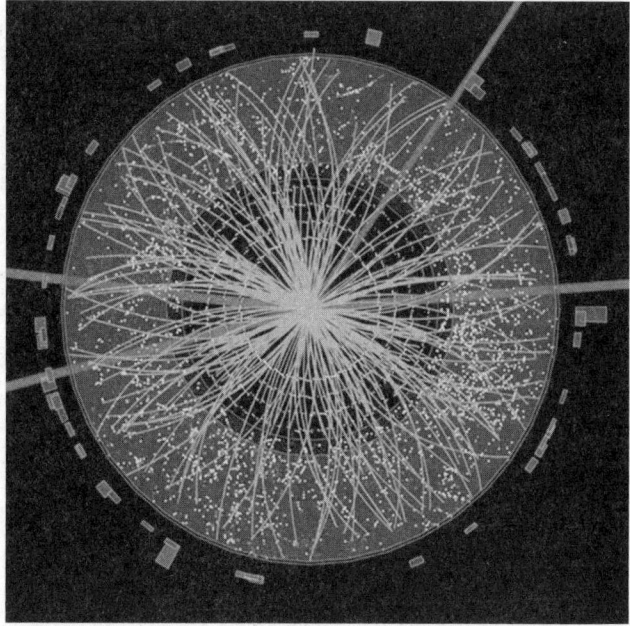

Figure 15.1. A compelling (to a physicist) image of a
Higgs particle exploding, taken at the CERN laboratory
near Geneva.

it has caught the public attention, and I even used it in the title for this
section.

The Higgs theory received a scientific accolade when Peter Higgs
and Francois Englert shared the 2012 Nobel Prize in Physics for their
prediction. Of course, the award was a minor one for Higgs himself,
compared with the immortality bestowed by having this key aspect of
physics named after him. Poor Englert had to be satisfied with only a
Nobel Prize.

The discovery of the Higgs is yet another blow to Eddington's claim
of a causal relationship between entropy and time. Here's how. In the
initial Big Bang, before the appearance of the Higgs, all particles were
massless. There is also good reason to believe that during this period,
even as the universe expanded, these massless particles had a "thermal"

distribution of energies; that is, they matched the kind of distribution you get when you maximize entropy.

Now, it had been known since the 1970s that such a collection of massless particles does not change its entropy as the universe expands. The key point is that in the early universe, the entropy of all the matter was in massless, thermalized particles, so it wasn't increasing. If time's arrow were truly being driven by the increase in entropy, there would have been no arrow. Time should have stopped. We never should have left that era. With stopped time, the expansion would stop (or never proceed in the first place). Without time, you would not be here now reading this book.

Time didn't stop. The universe expanded, the ylem of massless particles cooled, the Higgs field turned on through spontaneous symmetry breaking, and particles started behaving as if they had mass. And here we are.

Physicists have pondered the meaning of time in the very early universe (the first millionth of a second). Since space was so uniformly hot, they worry that there is no good way to find anything that could serve as a clock during that period. Because of the energy of the particles, and the high density, even radioactive decay would reverse. So how could time even be defined?

At the heart of that conundrum is the mistake of thinking that time flow was driven by entropy. That's backward.

How Did Eddington Fool Us?

Why was Eddington's entropy arrow argument so persuasive? I like E. F. Bozman's inadvertent explanation, given in the introduction he wrote for Eddington's 1928 book. He says that Eddington made his case through "exquisite analogy and gentle persuasion." That approach departs significantly from the usual experimental verification requirement for convincing physicists that a theory is right. Popper would not have been impressed.

Eddington (and virtually all popular authors on the subject) love to give examples of entropy increase. Drop a teacup and it will shatter into pieces. Play the movie backward and it will look wrong. Teacups don't build themselves. Yet we do have teacups. How did they get assembled? Instead of a movie with a breaking teacup, show instead a movie of a teacup factory, and you'll get the opposite impression. Humans construct teacups. Humans have organized and arranged and taken raw materials that have high entropy and refined the material, put the components together, and made the teacups. Without that manufacture, there would be no low-entropy teacup to smash. Play this movie backward, showing a teacup turning back into clay and water, and the reversed time would be evident.

We are surrounded by examples of entropy decreasing. We write books; we build houses; we create cities; we learn. Crystals grow. Trees take raw carbon dioxide (a trace gas in the atmosphere), selectively absorb it, separate and draw in water and dissolved minerals from the soil, and build magnificently organized structures. The entropy of a tree is vastly lower than the entropy of the gas and water and dissolved minerals from which it was made.

Humans take these magnificent low-entropy trees, cut them into boards, and construct buildings. If you watched a movie of a house being built, you would know the direction of time from the increasing order, not the increasing confusion; you would know it from the *decreasing* entropy. The arguments given by other authors who talk about smashed teacups are not general arguments; they have cherry-picked the examples that show entropy increase, when we are actually living in a world made better by the local decrease in entropy. (Cherry-picking itself is a form of local entropy decrease. So is writing a book.)

Of course, the entropy of the *universe* increases when we build a house. Most of the increase comes from the heat radiation that is thrown off into space. Locally, entropy decreases. Include the photons that fly off to infinity and total entropy increases.

Even in space we see entropy decreasing. Out of the primordial mixed-up soup of gases, particles, and plasma, a star is formed, and a planet around it, and life begins on that planet. The early Earth was originally a homogeneous mixed mess, hot and fluid; as it cooled, it differentiated, became more organized, putting the iron mostly in the core, rocks nearer the surface, and the gases in the atmosphere. It became enormously more organized, just as the cooling coffee cup lost entropy. Of course, in doing so, it ejected a lot of heat, which increased the entropy of the universe. That entropy was thrown off, mostly to infinity, while the entropy of the Earth decreased.

Watch the movie of the Earth's formation forward and backward, and it will be clear that the version showing entropy decreasing is the right one; you are watching the formation of structure on the Earth, not destruction to chaos. The history of the Earth from gas to liquid to solid, the history of life, the history of humanity, was not a history of increasing local entropy but of decreasing it. The history of civilization was not a history of breaking teacups but of making them.

Eddington also led us to believe that he was presenting science, but it did not really resemble the work of Newton, Maxwell, or Einstein. It was more along the lines of Augustine, Schopenhauer, and Nietzsche. It was philosophy, worthy philosophy, but it was not science.

Eddington's entropy arrow connection never was falsifiable. Worse, it never had empirical basis, nor has it developed any—not in the nearly nine decades since it was proposed. The only justification for it was that both entropy and time were increasing. That's a correlation, not causation. *Cum hoc ergo propter hoc.* How did Eddington fool us?

As Calvin says in the third panel of the strip shown in Figure 15.2, "What went wrong?! I thought this stuff was based on planets and stars! How could those be misread? . . . What kind of science *is* this?!"

Eddington didn't fool us. We fooled ourselves.

Figure 15.2. Calvin ponders, "What went wrong?! I thought this stuff was based on planets and stars! . . . What kind of science is this?"

16

Alternative Arrows

If entropy doesn't set the arrow of time, what does?

[A living organism] feeds on negative entropy, that is, it consumes order from its environment. . . . That change compensates for the entropy increase that it causes by living. . . . The trick by which an organism remains highly ordered is, in reality, by perpetually "vacuuming" order from its environment.

— *Erwin Schrödinger,* What Is Life?

Many alternatives to the entropy arrow of time have been proposed. These include the black-hole arrow, the time asymmetry arrow, the causality arrow, the radiation arrow, the psychological arrow, the quantum arrow, and the cosmological arrow. All are worthy of some discussion, although I find the last two in the list—quantum arrow and cosmological arrow—to be the most compelling.

The Decreasing-Entropy Arrow

The decreasing-entropy arrow might be considered a variant of Eddington's entropy arrow, although it is, in fact, fundamentally different in concept. Think of it as focusing not on the breaking of a teacup, but on the manufacturing that gave you one to break. This approach argues that time goes forward because space is empty and cold, so we can dump excess entropy into space like trash and forget it, enabling us to reduce our local entropy. In the decreasing-entropy arrow, local reduction determines the direction of time.

For the decreasing entropy arrow, I am making the implicit assumption that memory requires decreased entropy—that is, that memory requires the brain to become more organized rather than less, so we replace random connections between neurons with organized ones, neurons that can bring back details of past events and past deductions. As I discussed in the previous chapter, reduction of entropy is a key facet of the creation of life and of civilization. Schrödinger spoke to this question in his book *What Is Life?*, from which I quoted at the beginning of this chapter.

What makes the flow of time uniform? The evidence for uniform flow comes from the fact that when we look at distant events, their rate of time tends to conform to ours. If time flowed in fits and spurts, when we looked at distant events the fits and spurts would not line up with our current ones, and we would see irregularities. However, time could be gradually accelerating (or decelerating), and such a subtle change would not be noticed.

The issue we are addressing here is not the pace of time, but the arrow, the direction in which memories are created. Our own psychological experience is generated by the formation of memories. We tend to experience time in fundamental units. Movies with twenty-four still images per second tend to be blended by the brain into apparent continuous motion. It is quite different for a fly, which lives in a millisecond world. Then there are the mobile trees of Tolkien, the Ents (no relation to Ent-ropy) who consider days rather than milliseconds to be the natural units of time.

The decreasing-entropy arrow suffers from many of the same problems as the standard Eddington theory. Local entropy increases during the day (because of the temperature rise; hot things are less organized than cool things) and then decreases at night. Yet our experiential time continues to move forward. Is there some kind of time flywheel that averages out the short-term variations and gives us uniform progression of time? That has been proposed, but it is an ad hoc addition without falsifiability.

Perhaps we can get around this problem by focusing on the important entropy, the entropy of our minds, and ignoring the entropy of the biosphere as irrelevant. And I don't mean the total physical entropy of our brains, determined mostly by the temperature. I mean the entropy of thought, of memory, of organization and reproduction.

Mind entropy is virtually impossible to define, although we can try by using the methods developed originally by Claude Shannon for the entropy of information. Indeed, a great deal of work has been done on this in recent years, and the field is called *information theory*. Entropy of information has much in common with entropy of the physical world, and they share many theorems. Entropy of information also has paradoxes. How much information is stored in the number 3.1415926535. . . ? Is it infinite, or no more than is stored in the symbol π?

Despite these commonalities, I consider an information entropy model for the arrow of time to be far more plausible than Eddington's physical entropy theory. What we have not succeeded in doing is even estimating the information entropy in the human brain, and whether it is actually increasing or decreasing with time. (If it is turning a set of zero bits into a mixture of ones and zeros, then arguably memory is an *increase* in entropy.) Our memories certainly reorganize, and we try very hard to learn what is important, but nobody has yet devised a good measure of important information, and that would likely be the key to making such a theory viable.

The Black-Hole Arrow

Many objects in our universe are widely thought to be black holes already existing or black holes nearly formed. These include objects that are "small," meaning in this context only several times heavier than the sun (only in astronomy would that be considered small), and some pretty large ones: massive black holes in the centers of galaxies, weighing in (using that term metaphorically) at between a million and a billion solar masses.

Drop something into a black hole and it will not come out, ever again. Things fall in, not out. The recent theoretical prediction of black-hole radiation does not change this asymmetry; for the larger black holes that we are talking about, such radiation is so tiny as to be negligible, and the radiation doesn't actually come from the surface of the black hole, but from just above it. So you can determine time's arrow by watching things fall into black holes.

For many years, Stephen Hawking believed that objects falling into black holes violated the Second Law of Thermodynamics. That's because the object falling in essentially disappears from the universe, carrying its entropy with it, and making the entropy of the universe appear to decrease. I never found that argument compelling; we hardly need black holes, since sending a photon to infinity also eliminates entropy from an observable universe. (You can never catch up with a such a photon.) Eventually, Hawking changed his mind; his student Jacob Bekenstein convinced him that black holes do contain entropy themselves, and that when an object falls in, the entropy of the black hole increases; thus (when you include this component), the entropy of the universe does indeed go up, and the Second Law is saved.

What about the black-hole arrow? It doesn't really stand up to careful analysis. The fundamental reason is that, measured in the frame outside of the black hole, such as that of the Earth, the object never does reach the black hole. I discussed this in Chapter 7, "To Infinity and Beyond." So, actually, within any finite time interval (measured in the Earth frame), the object falling into the black hole can still, in principle, escape.

The possibility of escape is made formal by the postulated existence of "white holes." A white hole is a time-reversed black hole. According to the equations of general relativity, such objects could indeed exist. Do they? Not as far as we know. But their possibility shows that the equations of black holes show no intrinsic time asymmetry—not in our proper outside frame, and that's the frame in which the arrow of time is a mystery.

The Radiation Arrow

A quirk in the classical theory of electromagnetism led to a dispute between Walter Ritz, an eminent Swiss physicist, and Albert Einstein in the early 1900s. It was based on the known fact that shaking an electron causes the emission of electromagnetic waves. That's what we do with a radio antenna: we make electrons oscillate back and forth along a length of wire, and in doing so, they emit radio waves. On a smaller scale, any object that is hot (such as a tungsten filament in a lightbulb) is full of hot electrons, and hot electrons vibrate at high frequency, accounting for why they glow red or white-hot. The shaking electrons generate the high-frequency electromagnetic waves that we call visible light.

The emission of such radiation could be calculated using the classical equations derived by Maxwell, but doing so seemed to require an assumption about the direction of time. That's where the idea that radiation could drive time forward originated. In a modern junior- and senior-level physics textbook on electromagnetism, look at the section on radiation. The equation that describes radiation is named after the person who first derived it in 1897, Irish physicist Joseph Larmor. The books claim that to derive the radiation equation requires invoking a principle of "causality"; that is, you must assume (according to most texts I've looked at) that the shaking precedes the radiation. Causality is explicitly invoked by inclusion of what is called the *retarded potential* and omission of the *advanced potential*. In other words, you have to assign a direction to the arrow of time in order to calculate the radiation of light or radio waves.

The fact that causality had to be invoked to derive the equations of radiation led many physicists to believe that the process of classical radiation, a phenomenon that is present throughout physics (not just in light but in water waves, sound waves, and earthquake waves) is responsible for the arrow of time. Indeed, in examples I've given illustrating the decrease of local entropy (such as in the manufacture of a teacup or building), the emitted radiation is what accounts for the entropy decrease, by carrying off enough entropy to more than make up for it. So, radiation sets the arrow.

Ritz felt that the equations of electromagnetism, particularly the clear example of how radiation is calculated, did indeed have a built-in direction for time; Einstein argued that they didn't. It seems odd that there could be a dispute over a mathematical issue; the problem, however, was not in the math, but in the way the math was interpreted. Their dispute became public; it was carried on in a series of letters in the prominent journal *Physikalische Zeitschrift*. Finally, the journal editor asked the two physicists to submit a joint letter articulating their dispute. They wrote the letter, considered to be their "agreement to disagree." The dispute was over inclusion of the advanced potential, the part of the equations that seemed to give foreknowledge to the radiation of what the shaking electron was about to do. Ritz said that was unphysical; Einstein argued that, as part of the theory, it should be included.

When I consider their dispute in hindsight, it appears to me that Ritz was being driven by the conclusions he wanted to reach rather than by a compelling fact in the math. He had not yet been convinced that Einstein's still new relativity theory was correct, and Einstein's name had not yet become synonymous with genius; that was still a few years off. Einstein remained objective. It seems strange that Einstein didn't work out the math, but that remained undone until a young student named Richard Feynman presented his work to Einstein in 1945.

Feynman Advances

In 1945, Richard Feynman had just returned from the Manhattan (atom bomb) Project, where (with no PhD yet) he had been a very junior scientist. He claimed to be the only person to have disobeyed orders by keeping his eyes open to watch the moment of the first atomic bomb explosion in New Mexico (although he did, of course, watch through a dark filter). Feynman's thesis adviser at Princeton, John Wheeler, asked Feynman to study the asymmetry in the derivation of the radiation equation, and to see whether radiation could be derived using the advanced potential as well as the retarded potential. This can be thought of as asking whether knowledge of the future can be used to predict the past. Do the equations of classical radiation demand that time move forward, or could even radiation work backward?

Feynman was able to demonstrate that, indeed, the equations worked with both the advanced potential and the retarded potential included—a result that supported Einstein's position. He showed that the equations for radiation were symmetric in time; there was no intrinsic arrow. This demonstration was a remarkable achievement for a young graduate student, and a precursor of the great things Feynman was to do—including his later reinvention of quantum physics, and his interpretation of antimatter as matter moving backward in time.

Wheeler was delighted with Feynman's results, and he asked Feynman to describe his work at the weekly seminar organized by Eugene Wigner, the physicist whose mathematical genius set the foundation for much of modern theoretical physics. It was Feynman's first scientific talk, and he was intimidated by the prospect of lecturing to Wigner, but he agreed to do it. Then Wheeler told Feynman that he had also invited Henry Norris Russell, famous for his contributions to both the theory of stars and the theory of atoms. Feynman became even more nervous. Then Wheeler invited John von Neumann, one of the outstanding geniuses of any era, a man who made seminal contributions not only to physics and math, but also to statistics, digital computer theory, and economics. Making it much worse, Wheeler then invited Wolfgang Pauli, a cofounder of modern physics, one of the most intimidating of the great physicists of the quantum era, and the discoverer of the Pauli exclusion principle that accounted for the stability of atoms. Pauli was known for his sharp and unforgiving criticism of work that he considered inferior. Feynman thought it couldn't get any worse.

Yes it could. Einstein accepted an invitation to come.

Feynman said he was totally intimidated. "Here are these monster minds in front of me," he recalls in his book *Surely You're Joking, Mr. Feynman!* Wheeler tried to comfort him with these less-than-reassuring words: "Don't worry; I'll answer all the questions."

Feynman said that when he finally started talking, all his nervousness suddenly vanished. He lost himself in pure physics, and he discovered that the expert on the subject was not Wigner, not von Neumann, not Pauli, not even Einstein, but Richard Feynman. He, not Wheeler, answered the questions, and everything worked out fine.

Feynman showed that classical radiation theory does not distinguish past from future. Einstein was right, not Ritz. (Surprised?) Electromagnetic radiation does not define the arrow of time.

Figure 16.1. Richard Feynman, appearing on a US postage stamp honoring him and his work. Several "Feynman diagrams" appear in the background. The arrows on the positron lines (to the left on the stamp) indicate that the positrons are moving backward in time.

The Psychological Arrow

This is, in many ways, the most intriguing of the proposed arrows of time. If we make the assumption that physics is completely reversible in time, that the movie run backward violates no laws, then might there still be an arrow of time determined by life? Does something cause us to remember the past, and not the future, even though the laws of physics are symmetric?

Most physicists believe that there is nothing spiritual about the direction of time, that it is not related to any special nature of life, but that the answer lies completely within physics. Stephen Hawking, for example, asserts that the psychological arrow is based on the entropy arrow. But that is a tricky conclusion to make. It is not usually argued, but simply stated, as self-evident. Hawking says, "Disorder increases with time because we measure time in the direction in which disorder increases. You can't have a safer bet than that!" This statement is an example of the logical fallacy known as *ipse dixit*, proof by emphatic assertion, by authority.

What is memory? It turns out to be harder to define, to understand, than you might expect. We all have the sense that when we learn something, we are decreasing the disorder of our brains. Is that a decrease in entropy? It can also be seen as an increase—if our brains are simply very well organized blank slates (like computer memory that is full of zeros with no ones) and when we learn, we are making them more complex, more "disordered" in an information sense. But it is generally accepted that, if memory is a decrease in disorder, the process of learning generates plenty of heat that increases the entropy of the universe but is discarded. So even though the local entropy of our brains decreases, the entropy of the universe as a whole increases. It's the local entropy decrease that most matters to us.

Some people think that life, that consciousness, is a phenomenon transcending physics. I'll discuss that possibility later in this book. To the extent that we consider a human to be a big, complex combination of chemicals responding to external stimuli, no psychological arrow need be postulated. We just go with the flow—and the creation of new entropy. Computers, running purely on physics equations, are perfectly able to remember the past, with no need for "psychology" or consciousness or life, and they have decided difficulty in deducing all but minor aspects of the future. In this picture, in the tachyon murder paradox, Mary had no choice but to pull the trigger; her free will was an illusion, and her behavior was completely determined by physics equations.

The Anthropic Arrow

Anthropic means "relating to humans." The earliest use of the term recorded in the *Oxford English Dictionary* (1859) refers to the observations of gorillas and their humanlike behavior. The *anthropic principle* is beloved by many present-day theorists, particularly in string theory. It states that we can determine the parameters of the universe, such as its age and size and makeup, and maybe even the direction of time, by the fact that only a very narrow range of possibilities could lead to intelligent life.

According to the anthropic principle, the very fact that we are thinking about the origin of the universe today could only happen if our universe was a very special one. I think, therefore I am, and moreover, that means time must move forward and not backward. Hawking says the anthropic principle is so powerful that it even determines why our psychological arrow of time points in the same direction as the entropy arrow. If it didn't, he states, we wouldn't be discussing the issue. QED.

I personally regard the anthropic principle as useless. In my experience, it is used as an excuse by physicists who have failed at computing something, so they argue that things must be the way they are because otherwise we wouldn't be here discussing the issue. Such reasoning depends on the assumption that any form of intelligent life must be very similar to our own. Time must go forward because reality would be different if it went backward.

A colleague of mine, Holger Müller (no relation that we know of) suggested an example that illustrates the emptiness of the anthropic principle. Imagine scientists pondering, "Why does the sun exist?" The anthropic answer would be: "Because if it didn't, we wouldn't be here!" That's a simplistic answer that might have come from eighteenth-century philosophers. The much more fulfilling and rewarding answer has come from physics: A cloud of debris that was left over from a prior supernova explosion self-gravitated. As the pieces fell in, their velocity of motion and gravitational compression turned into heat, raising the temperature enough to ignite thermonuclear reactions. And so on. That's the kind of answer that fits into the paradigm of science, well beyond the hollow approach of the anthropic principle.

In the early 1900s, Wolfgang Pauli, one of the founders of quantum theory, was shown a paper that he considered sloppy and confused. He is purported to have said that the work "was not even wrong." In his mind, one of the virtues of a scientific theory is that it could be falsified; the theory he was shown failed to meet this criterion. Peter Woit, a mathematical physicist at Columbia, has argued eloquently that the anthropic principle (as well as string theory) also fits Pauli's measure of being "not even wrong." He lays out this argument in his blog and book, both titled (naturally) *Not Even Wrong*. In my mind, the expression *not even wrong* equally describes the entropy explanation for the arrow of time.

Time Reversal Violation

In studying *now*, we will soon enter the realm of quantum physics, the *other* theoretical revolution of the twentieth century (in addition to relativity). Some key concepts of quantum physics are just as bothersome as the most disturbing issues in relativity, such as the loss of simultaneity and the reversal of events—perhaps even more so. They include particles going backward in time (antimatter), and the mysterious phenomenon called *measurement* that appears to have its own arrow. But before I discuss those topics, I need to talk about a quantum effect that is more immediately related to the arrow of time. Discovered in 2012, it is called *time reversal violation* or *violation of T symmetry.*

Time reversal violation means that a movie of elementary particle interactions could be determined, unambiguously, to be running forward or backward. There really is a direction of time built into the microscopic world of elementary particles that is completely independent of any entropy linkage. The discovery of such a process was a long-term goal (I refuse to use the overworked metaphor *holy grail*), extremely slow in coming, very challenging experimentally, and a great achievement. The violation of T had been long suspected, because of prior observation of important differences in the behavior of particles and antiparticles that seemed to imply that T violation should also be expected.

As a graduate student in the 1960s, I worked on particle interactions with Phil Dauber, who was then a recent hire in the group led by my mentor Luis Alvarez at the Lawrence Berkeley Laboratory. I got very excited when one of the particles we were studying, called a *cascade hyperon*, appeared to show T violation in its decay! Because such a discovery would be so important, Phil worked intensely on the data, doing every test he could imagine, looking hard for possible systematic biases, doing his best to falsify the discovery.

Finally, he told me he had managed to reduce the observed T violation down to two standard deviations, meaning that it had "only" a 95 percent chance of being right, a 95 percent chance of truly showing time reversal violation, and a 5 percent chance of being wrong. Those odds, he explained, are not good enough for a major discovery. They give you a 5 percent chance of publishing something that is complete nonsense. I was upset. I thought 95 percent chance of such a great discovery was pretty good odds. No, it isn't, Dauber patiently explained to me. We in particle physics, he said, have high standards. He wrote most of the paper (I was a coauthor), pointing out that the parameter indicating T violation was only two standard deviations away from zero and therefore (in his words) was statistically consistent with being zero. We reported no violation of T. There were no newspaper headlines.

Imagine my disappointment. I had joined a project that had made one of the most important potential discoveries of all time, something my descendants would someday read about in their history books, and it had a 95 percent chance of being right! But Phil simply wasn't convinced that a 95 percent chance of being right was good enough.

Decades later I checked back. Over time, more precise measurements of the T parameter for the cascade hyperon had been made, and interestingly, the proper final value was indeed zero, within much smaller error uncertainties than we had been able to achieve. Phil had been absolutely right in his strict scientific standards, and I learned a very important lesson about discovery.

What had gone wrong? How could our discovery have a 95 percent chance of being right and still turn out to be wrong? The answer came from the fact that we were studying a huge number of different phenomena. We were looking at the decay of different kinds of particles, at their interactions, at their masses and expected symmetries. In our paper we were reporting more than twenty different new results. If each had a 5 percent chance of being wrong, then indeed we should expect one out of twenty to be wrong. The only way to avoid serious mistakes is to hold high standards.

When I think about the history of the Alvarez group, I realize that I had the good fortune to work with an amazing collection of some of the top physicists in the world. From the 1960s through the early 1970s, they were on the forefront of particle physics and were reporting new discoveries virtually every month. It might very well be true that the number of important discoveries they reported exceeded that of any other physics group in history. Yet I cannot find a single instance of anything they reported that was later shown to be wrong. That is an amazing record. Achieving it required high standards.

In 2012, a group at the Stanford Linear Accelerator Center published results of a study of two different reactions having to do with the radioactive decay of a rare particle called a B. The B comes in several forms, including one called a $\bar{B}^0$ (pronounced "bee zero bar") and one called a

B_- ("bee minus"). They studied two reactions: one in which a $\bar{B}^0$ turns into a B_-, and another in which the exact opposite happens, with a B_- turning into a $\bar{B}^0$. These are time-reversed reactions; if you saw a movie of one, it could be a movie of the other one being played backward. But in studying the two reactions, the group observed a departure from symmetry that amounted to 14 standard deviations. According to statistics theory, such a result has only one chance in 10^{44} of being wrong. That's one chance in a hundred tredecillion. That certainly would have been enough to satisfy even Phil Dauber.

It was not a serendipitous discovery. There were very good reasons to look at these particular reactions, based on prior observations of peculiar behavior in related particles called kaons. The researchers were looking for, hoping to see, time reversal violation. We can now say clearly something that we could only speculate on before 2012: time reversal is not a perfect symmetry of the laws of quantum physics. Forward time is different from backward time, in the heart of physics itself.

This is a very important insight in our study of the nature of time. But can this effect play a role in determining the arrow of time, its flow, or the meaning of *now*? I think not. Time reversal violation is a tiny effect. Using a metaphor, the law of time invariance is broken, but it is hardly a misdemeanor, no more than a parking ticket, and certainly not a felony. The only evidence we have comes from a special kind of radioactivity (*B* decay) that is seen only in exotic high-energy physics laboratories. How could such a small, difficult-to-observe phenomenon play a role in setting the direction of time?

These statements suggest to me that time reversal violation does not play a role in our current experience of time. That does not mean it was unimportant in the early moments of the universe, when all of space was filled with a thick, hot ylem of particles, including (in the extremely early universe) lots of kaons and *B* particles.

In fact, an argument has been made to show that the closely related matter-antimatter symmetry violation may have been central to the creation of the universe as we know it. Andrei Sakharov, the physicist behind the creation of the Soviet hydrogen bomb, and a Nobel Peace Prize winner (for his courage in defying the Soviet government), pointed out in 1967 that such a violation of matter-antimatter symmetry, called "CP symmetry," could have caused a tiny excess of matter over antimatter in the early moments of creation, only one part in 10 million. But then, in the weird early moments of the universe, as it cooled, all of the antimatter annihilated with matter, turning into photons. Because of the slight excess, however, there was a little left over—the stuff we call *matter*, which now makes up all of the matter of the universe. Stars, planets, people—all are made from that tiny bit of matter that was left over from the great annihilation. The violation of CP symmetry was small, and there was only that minuscule surfeit of matter, but vive la différence!

The observation of time reversal violation is deeply significant for another reason as well: the fact that it was predicted on the basis of fundamental aspects of quantum theory, an abstract result called the *CPT theorem*. The fact that this theorem predicted an unusual phenomenon and still was verified is yet another indication that quantum theory is on solid ground.

The Quantum Arrow

A time asymmetry may be lurking in a mysterious aspect of quantum physics known as *measurement*, a process that appears to affect the quantum states of the future but not of the past. The next few chapters will discuss this process in some detail. The main drawback of leaning on the theory of measurement is that it is so poorly understood that an explanation that depends on it isn't truly an explanation, but only a hope that two mysteries (time and measurement) can be reduced to one. Nevertheless, a quantum arrow needs to be seriously considered.

The Cosmological Arrow

Eddington proposed the entropy arrow because entropy increase was the only law of physics that appeared to have a direction in time. The question remained, why does entropy increase? The answer was found in the Big Bang, a fantastic discovery that could account for the fact that our universe is not yet totally confused. The Big Bang allowed the universe to be young—and therefore not yet completely randomized—and the expansion of space gave plenty of room for additional entropy increase.

But with the discovery of the Big Bang, we really should look at the issue of the arrow of time afresh. The entropy mechanism does not really work very well. Is it needed? If we think of the universe in terms of space-time, why should the universe be expanding only in terms of space? Why not also time? In fact, it obviously is; every second, we add a new second to time. Perhaps the flow of time is more accurately thought of as this creation of new time. Think not of a 3D Big Bang, but of a 4D Big Bang, with continual creation of both new space and new time.

In Chapter 11 I offered the following challenge: Suppose you were given complete, God-like knowledge of the universe for two instants of time and asked to figure out which instant came first. How would you do it? The answer I gave there was to calculate the entropy of the two snapshots. Whichever one had lower entropy came first. But you could also just look at the size of the universe. The smaller one came first.

To understand this fully, we need to delve into the other great and revolutionary discovery of the twentieth century—one that, in many ways, is even more disconcerting and counterintuitive than relativity. It is the perplexing reality of quantum physics.

PART III

SPOOKY PHYSICS

17

A Cat Both Dead and Alive

*Introducing quantum physics by starting with
the most absurd example . . .*

"I can't define [it] . . . but I know it when I see it."
— *Supreme Court Justice
Potter Stewart (not on
the subject of measurement)*

As if the mind-twisting concepts of relativity theory weren't disruptive enough for the twentieth century, another equally distressing yet equally important revolution took place immediately afterward: the development of quantum physics. One of its founders was Albert Einstein, who deduced that the energy of light is quantized, detectable only in bundles that we now call *photons*. But quantum physics didn't settle down as rapidly as relativity had. It contained features so strange and mysterious that the very inventors of the theory continued to argue and debate what it meant, how it should be interpreted, and whether it was just a temporary approximation with a more complete description of reality hiding underneath, yet to be discovered. That debate continues to this day.

The problem with the theory comes from its very formulation. Quantum physics postulates that the real world is described by something shadowy and ephemeral, unmeasurable even in principle, called an *amplitude*. An amplitude could be a single number, a complex number with both real and imaginary parts, or a collection of numbers called a *wave function*. Quantum physics postulates that the amplitude is wraithlike, unreachable, a background spirit that carries with it all of reality. Even if you know the amplitude exactly, however, you can't predict the result of a measurement exactly, but only the probability that a measurement will yield a particular result.

All this sounds mystical and tentative, yet these principles are used today to design the electronics that run our smartphones, tablets, TVs, digital cameras, and computers. Virtually every physicist works with the ghostlike amplitudes and wave functions. Most just ignore the unmeasurable aspects of quantum theory and get on with their jobs.

Not Einstein. All his breakthroughs in physics came from focusing on paradoxical results, unexplained phenomena, things that didn't make physical sense to him. The new quantum physics had aspects that fit right into these categories—more enigmatic than time dilation and length contraction, queerer than black holes, harder to imagine than the reversibility of events in time. Perhaps the most distressing aspect, even today, is exemplified by a story made up by Erwin Schrödinger, the physicist whose name is known to every physics student for the *Schrödinger equation*, the most important equation in elementary quantum physics. He was a colleague of and sympathizer with Einstein, and he shared Einstein's discomfort with quantum physics.

Schrödinger's Cat

Schrödinger devised a vivid example in support of Einstein's contention that quantum physics was fundamentally unsound. The setup is simple, although purposefully cruel, presumably to grab your attention and force you to appreciate the cognitive dissonance that the story provokes.

Schrödinger asked us to imagine a cat in a box. The box also contains a radioactive atom that has a 50 percent probability of undergoing decay in the next hour. If it decays, it will trigger a mechanism that kills the cat. Schrödinger vividly suggested a hammer breaking a vial of hydrocyanic acid. Want to see a depiction? Look up "Schrödinger's Cat" online.

Open the box in an hour and the odds are 50 percent that you'll find a dead cat, 50 percent that you'll find a live one. That seems pretty straightforward, if inhumane. (Don't try it at home.)

What is weird, what bothered Einstein and Schrödinger, is the way the situation is described in the language of quantum physics. According to the standard approach used by virtually all physicists, the amplitude that describes the atom and the cat evolves over the hour. Initially, that amplitude describes a live cat and unexploded atom. But as time passes, the amplitude changes. At the end of the hour, the evolved amplitude consists of two equal parts: one that has a dead cat with fragments from the exploded atom, superimposed on one that has a live cat and unexploded atom. Until someone peeks, the cat is simultaneously both dead

Figure 17.1. A film strip depiction of Schrödinger's cat. As time goes on, two quantum states develop: one with a live cat and one with a dead cat. Only when a human peeks into the box is one of these states randomly chosen to represent reality. (Illustration by Christian Schirm.)

and alive. According to the rules, the act of opening the box and looking in constitutes a *measurement*, and when measured, the wave function immediately *collapses* and you are left with only one reality, not a superposition of two. When observed, the cat is either fully alive or fully dead, no longer both. That simplification takes place only when you peek.

I asked my wife, Rosemary (an architect), to read this chapter. She found the section on Schrödinger's cat, up to here, completely incredible. She refused to believe that any scientist could seriously postulate both an alive and a dead cat simultaneously. That notion was so absurd, so ridiculous, that she stopped at this point and would not read the rest of the chapter until I corrected the stupid description that gave the impression that quantum physics included such nonsense.

But ask any physicist. This is the way it is. Take comfort in the fact that you are bothered by the same issue that bothered Einstein and Schrödinger, the same issue that inspired Schrödinger to pose this crazy example. That's what I told Rosemary, and she agreed to keep reading—although, perhaps, under protest. (And she gave me permission to describe her experience, as a comfort to others.)

Schrödinger and Einstein considered this story to be a *reductio ad absurdum*, a "reduction to an absurdity," a ridiculous conclusion demonstrating that quantum physics is preposterous and therefore false. Until you look, the cat is both dead and alive? Come on! In their minds, this example should have won the day, should have ended the discussion and demonstrated that quantum physics was fundamentally flawed.

Max Born and Werner Heisenberg, originators and supporters of the probability interpretation, refused to back down. Yes, the story of Schrödinger's cat sounds ridiculous, but so did time dilation and space contraction when Einstein first proposed them. Even the theory that ordinary matter consists of atoms once defied common sense. The cat story contains no contradiction—only a situation that goes against intuition.

That argument took place about eighty years ago. What is the situation today? The remarkable answer: virtually all physicists accept the Born-Heisenberg point of view. Yet the absurdity of Schrödinger's cat has never been satisfactorily answered. How do today's physicists respond to the *reductio ad absurdum*, the ridiculous example, of Schrödinger's cat? They don't. Schrödinger's cat still bothers them today, when they think about it, but then they choose to ignore the problem and move on.

The Copenhagen Interpretation

The approach of Born and Heisenberg (and they, too, were founders of quantum physics) has come to be called the *Copenhagen interpretation*, so named by Heisenberg after the city where he worked as an assistant to Niels Bohr. Most physicists today accept the Copenhagen interpretation. Einstein continued to dispute it up to his death in 1955. Meetings are still held in which the few and the proud debate the reality of quantum physics, with long mathematical and esoteric discussions of possible alternatives, but most physicists ignore those meetings. Quantum physics works; that's good enough for the silent majority of physicists. Ask one and you'll likely get a reply like this: "I know it sounds weird, but there is no way we could tell whether the cat was alive or dead without affecting the outcome, so we are unable to tell the difference."

Some scientists misunderstand quantum physics and mistakenly believe that the cat is either dead or alive, not both, and that the observer just doesn't know until the box is opened. That's what Einstein and Schrödinger thought. That approach is now called *hidden-variable theory*. In this case, the hidden variable is the livingness of the cat. That's the way it is frequently taught in undergraduate courses, but that is not what the Copenhagen interpretation says. And as I'll show, experiments in a feature of quantum physics known as entanglement imply that the Copenhagen interpretation is the correct one, not the Einstein-Schrödinger hidden-variable view. In Chapter 19 I will describe the first such experiment, which was conducted by Stuart Freedman and John Clauser. (No, they didn't use a cat.) The best theory we have holds that, indeed, the Copenhagen interpretation is the right one; the cat is both dead *and* alive until the moment of measurement.

Couldn't you tell whether the cat had died earlier by the state of the body, the temperature of the blood, or some other physiological sign? Actually, the amplitudes of the atom and the cat will include all possible decay times, suitably weighted to reflect the probability of an early versus late radioactive decay. (If you include that additional aspect in your measurement, then the amplitude will be a bit more complicated than a single number.) If you peek or insert a thermometer, that counts as a measurement. When you open the box, you may see a freshly killed cat, or one that looks like it has been lying dead for almost an hour, even though, according to the Copenhagen interpretation, a moment earlier its fate had not yet been decided.

Couldn't the cat tell? What do we mean by a *measurement*? Does it require a person, or can a cat make a measurement? What if we replaced the cat with a person? As amazing and disturbing as this will sound, the answer to all these questions is *we don't know*. A true theory of measurement doesn't yet exist. It is a dream in the minds of physicists. And this not-yet-formulated theory of measurement is where some physicists believe the origin of time, its arrow, and its pace could lie. When you peek, you affect only the future amplitude; the future holds a cat either alive or dead. You don't affect the past amplitude, which consisted of a cat both alive and dead. Thus there is an asymmetry, something new in physics that distinguishes the past from the future.

The Ghost Underlying Reality

For Schrödinger's cat, the dead/alive amplitude was simply a number that, when squared, gave the probability at the end of a period of time. As I mentioned earlier, if the amplitude depends on location and time, it is called a wave function. Schrödinger, the originator of the cat story, is most famous for having worked out an equation that shows how the wave function responds to outside forces, how it moves and varies in space and time, the famous *Schrödinger equation*, studied by all physics and chemistry students.

"About your cat, Mr. Schrödinger—I have good news and bad news."

Figure 17.2. Erwin Schrödinger in *The New Yorker*.

A wave function can describe an electron moving through space or orbiting an atom. In chemistry, the wave function is called an *orbital*. Because wave functions are not point-like but are spread out, the position of the particle (where it will be detected) is uncertain. The velocity of the particle, defined by the pattern of the wave function, is also uncertain. All wave functions vary with time, and the energy of a particle is directly related to the frequency, by the same formula that Einstein discovered for photons, $E = hf$. If the frequency is not precise, if the pattern of vibration is like that of a musical chord (containing many notes), or worse, like a noise, then the energy, too, is uncertain.

To find the expected position of a particle, square the numerical value of the wave function everywhere. That will give you the relative probability of finding the particle at any location. Do a wavelength analysis to determine how fast the particle is moving. Short wavelengths correspond to high velocity. French physicist Louis de Broglie showed that the momentum of a wave function (mass times velocity) is given by Planck's constant h divided by the wavelength: h/L.

In some cases, the wave function can be a complicated superposition of complex numbers. When you make a measurement, the wave function will "collapse," changing to something in agreement with your measurement. That change is called a *collapse* because it typically simplifies the wave function. Open the box to look at Schrödinger's cat and the wave function collapses to represent either a live cat or a dead cat but not both. All we ever see are the simple results of measurements, and these don't contain weird combinations like dead and alive—just dead *or* alive.

This wave function is truly ghostlike. It can't be measured. Every point on the function consists of typically two numbers (real and imaginary parts)—more if there is a superposition. Make a measurement and the new wave function is much simpler. This was part of the Born-Heisenberg Copenhagen concept, and it is still accepted today. In fact, today physicists are trying to take advantage of the hidden ghostlike aspects of the wave function by using them in quantum computers. In computer lingo, an amplitude bit is called a *quantum bit*, or *qubit*.[*]

The wave function of an electron could be small in extent, orbiting a nucleus, or large, filling space between Earth and the sun. If you know its past and the forces acting on it, you can determine (for example, by using Schrödinger's equation) what the wave function will be like in the future, but you can't examine the wave function with instruments without causing it to change, to collapse. When you measure the position of an electron, the new collapsed wave function might be highly localized or spread out with a size given by your measurement uncertainty.

What does it take to make the wave function collapse? We don't know. I'm serious. When physicists don't understand something, they

[*]   Technically, a qubit has two *states* available, whereas a bit has two values (0 or 1) available.

frequently give it a name just to be able to talk about the puzzle. In this case, the thing that makes the wave function collapse is a *measurement*. As I just said, we don't know what we mean by that. Generally, physicists ignore the problem and fall back on the famous Potter Stewart line: "I can't define [it] . . . but I know it when I see it." But the fact is that we don't really know it when we see it. Some people argue that it requires some sort of "consciousness." That's unhelpful, since we have no good understanding of consciousness. Einstein mocked this claim when he wryly commented, "Do you really think the Moon isn't there until we look at it?"

It wasn't just cats in boxes that bothered Einstein.

Quantum Theory Violates Relativity

You don't have to kill a cat to run into quantum paradoxes. Imagine an electron described by a very large wave function, extending from here to the sun. Detect that electron and the wave function will collapse, immediately and instantly, into a new wave function no larger than your detection apparatus. We knew we had one electron, and we know it is now at the Earth. We therefore know that it is not now at the sun. The theory says it collapses instantaneously. Is that compatible with our understanding of relativity?

I used the word *instantaneously*, but its meaning depends on the frame of reference. According to relativity, the two separated events (detection at the Earth; disappearance of the wave function at the sun) will not be simultaneous in all reference frames, even if it is simultaneous in the proper reference frame of the detection apparatus. That means there is a frame in which the disappearance of the wave function preceded the measurement. Moreover, there is a reference frame in which the wave function stayed around for a while afterward. Thus, according to the rules of quantum physics, there is a reference frame in which the electron, detected at the Earth, will still have a nonzero value at the sun. That means there is still a chance that the electron will be detected there.

But that's impossible. The electron has already been detected at the Earth. And there was only one electron. (Yes, we can set this up in such a way that we are sure only one electron is present.) Something is wrong.

An obvious explanation is that the electron isn't really an extended object but is point-like, and the wave function only expresses our ignorance about where it really is. That is the way quantum physics is often taught, and it is the way many working physicists think about it, but it is wrong. The idea that there is a greater reality and quantum physics simply describes our ignorance is just hidden-variable theory, with the *actual* but unknown position of the electron the hidden variable. Experimental tests have been performed to see which theory is right. In all of the experiments so far, quantum theory has won and hidden-variable theory has been falsified.

That means that the wave function does not obey relativity. That's rather disturbing; relativity has been extensively tested, with numerous experiments over the past century. How can we resolve this conflict between relativity and quantum physics?

18

Tickling the Quantum Ghost

*The mysterious issue of measurement, and how poorly
we probe the quantum wave function . . .*

[It's] like a box of chocolates. You never know what you're gonna get.
— *Forrest Gump*

Wave functions have many properties that make the analogy to
ghosts seem more than a metaphor. As we've already discussed,
their collapse is not limited by the speed of light. Therefore, in certain
reference frames, their collapse will move backward in time. The wave
function's only connection to reality is when we probe it, when we try
to measure the position or the energy of the particle it represents. When
we do that, according to quantum physics, the wave function changes in
ways that violate our intuition and seems to disobey our understanding
of relativity.

Are you shocked that modern physics would contain such a beast?
One of the founding fathers of quantum physics, Niels Bohr, said,
"Anyone who is not shocked by quantum theory has not understood
a single word." Richard Feynman said, "It is safe to say that nobody
understands quantum mechanics."* John Wheeler, Feynman's men-

* Feynman was using the older term quantum *mechanics*. When first formulated,
quantum physics did indeed focus on issues of mechanics, objects and how they move,
but modern quantum physics also covers the quantum behavior of fields, including

tor, also a key player in the way quantum physics developed, said, "If you are not completely confused by quantum mechanics, you do not understand it." Roger Penrose, one of the leading modern-day thinkers on the meaning of quantum physics, wrote, "Quantum mechanics makes absolutely no sense."

This crazy, impossible-to-understand theory, quantum physics, despite its ghostlike and confusing nature, is at the heart of all modern physics. It may be ethereal, but it gives precise and accurate predictions. Simply ignore the ghostlike aspects, learn how to solve the equations, and you can compute the future with remarkable (but not comprehensive) accuracy.

The equations of quantum physics, such as the Schrödinger equation, enable you to calculate how the wave function of, say, an electron, will change when you put a force on it. But the wave function isn't really the electron. It is the amplitude, the spirit of the electron, its apparition, its soul. We can never detect or measure the wave function. We can only calculate it, or probe its value at a spot. But when we make that probe, when we conduct a measurement, we change the wave function forever, immediately, irrevocably, and instantly.

Pwaves and Wavicles

Suppose you put a measurement device in front of the electron wave function—for example, a wire that senses electric current. If the electron wave function is wide, then only part of it will hit the wire. That means there is only a small chance that the electron will be detected. Using the wave function and the size of the wire, you can calculate the likelihood that the electron will hit it and be measured.

As the electron moves, the wave function behaves as a wave—hence its name. You can send the wave of a single electron through two dif-

---

electromagnetic and nuclear fields, and as such I think it's clearer to use the modern term quantum *physics*.

ferent and separate paths simultaneously, just as a single sound wave can travel to both of your ears. But when the electron is detected, it is detected as a burst, a sudden impact, a *quantum*. In many ways it seems to be a particle.

Which is it—particle or wave? The correct answer is that it is neither. We can understand the electron only in terms of a new construct, something we might call a particle wave, or a wave particle, or something else. Several times I've had my students vote on whether it should be called a *wavicle* or a *pwave*. Neither name predominated in the votes. It is not a wave; it is not a particle; it has some properties of each, but the mixture is very strange. It moves through space like a wave; it responds to measurement like a particle; it is a wave that can carry mass and electric charge. It can spread, reflect, and cancel itself, just as noise-canceling earphones cancel sound waves. But when you detect it, the event is typically sudden, abrupt. The detected electron continues to exist, but the wave function has been irrevocably altered. If you detect it with a small instrument, the previously large wave function instantly becomes small.

Going down the Drain

Einstein himself was the first person to suggest the particle-wave duality, in his 1905 paper on the photoelectric effect that described how light knocks an electron off a piece of metal. He suggested that light is indeed a wave, but when it is detected, when it ejects the electron from the surface, it always does so in a burst—a behavior reminiscent of a particle, not a wave. Sometimes it does this quickly, before a classical electromagnetic wave could have delivered sufficient energy. As noted earlier, Einstein said that the energy of the light quantum would be related to the frequency of the wave by the equation $E = hf$, where h is Planck's constant, the number that Planck deduced when investigating the glow of hot objects.

Einstein never imagined that the same equation would apply to electrons. Louis de Broglie made that suggestion in his PhD thesis in 1924. That was the breakthrough that ignited the rapid development of quantum physics. Electrons and photons were recognized, thanks to de Broglie, as being very much alike; the differences that had once been seen as central (one had zero rest mass; one had an electric charge) became secondary. They were both just quantum particle waves (pwaves? wavicles?). It was a great unification of physics.

Within three years, Schrödinger, Born, Heisenberg, and others had worked out the equations for how the waves respond to forces. Then Dirac showed how to reconcile the equation for the electron with relativity theory (although he didn't address the measurement conundrum); he devised a *relativistic wave equation* for it. The 1920s were a period of unbelievably rapid development. It was dazzling even to the physicists themselves.

The ghostlike feel of quantum physics bothered many physicists back then, and it still does. It typically takes students in physics and chemistry several years to get used to it. Physicist-mathematician Freeman Dyson once told me that there are three stages the student goes through in getting used to quantum physics. In the first stage, the student wonders, how could it be so? In the second stage, the student learns how to do all the mathematical manipulations and discovers the power of quantum physics calculations. The math predicts outcomes of experiments with amazing accuracy. The final stage, according to Dyson, is when the student no longer remembers that the subject originally seemed so mysterious.*

Not all physicists reach that last stage of contentment. The great successor to Einstein, in my mind, was Richard Feynman. More than anyone else in the twentieth century (with the possible exception of Enrico Fermi), Feynman had a deep intuition that led him to extraordinary insights and discoveries in diverse aspects of physics. But Feynman stayed away from the "interpretation" of quantum physics. In his colorful Brooklyn colloquial manner, Feynman warned, "Do not keep

---

* These are analogous to the three stages for establishing truth, attributed to Schopenhauer. See the quote at the beginning of Chapter 4.

saying to yourself, if you can possibly avoid it, 'But how can it be like that?' because you will go 'down the drain,' into a blind alley from which nobody has escaped."

Intrinsic Uncertainty

A key feature of the new quantum physics was a discovery that still bothers students and professors today. It is called the *Heisenberg uncertainty principle.*

Just the process of attributing wavelike properties to electrons causes immediate problems with our classical understanding. Let's think about ordinary waves, water waves. Such waves don't have precise positions; they are spread out. You may find it more surprising that many water waves don't have a precise velocity. Throw a rock into a moderately deep pool and watch the waves get wider. What is the velocity of the waves? You may think you know, as you look at a single crest and watch it move. But then you'll see that the crest disappears; the wave is still there, but the crest you chose is gone! A new one has appeared to replace it, but it appeared behind the one you were watching. It is clearly the same wave; it is there only because you threw in the rock.

In looking at waves, physicists recognize that a wave, such as one generated by a rock splashing in water or by a boat moving by, typically consists of a group of crests and troughs. For water waves, the individual crests move at a different velocity from that of the group. For deep water, the crest velocity (sometimes called the *phase velocity*) is twice the group velocity. Which one is the velocity of the wave? For quantum physics, the velocity of the group is what's important if you want to detect the particle far from its source.

Perhaps even more confusing, the group becomes broader as the wave progresses. It might have been quite narrow when it started, but by the time it has moved a great distance, it is much wider. What is the velocity of the wave—the velocity of the crests, the velocity of the front of the group, the velocity of the rear of the group, or the average? You can

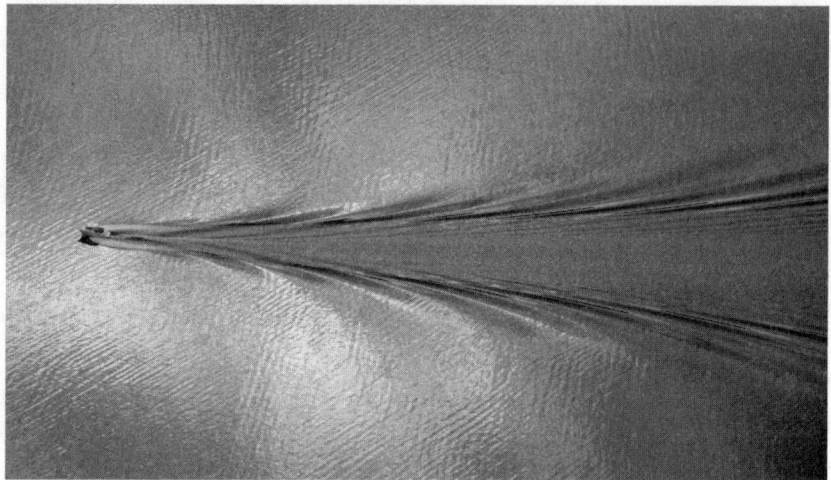

Figure 18.1. Water waves behind a boat.

see all these effects (crests appearing and disappearing, a group of waves getting wider) in the aerial photo in Figure 18.1, which shows waves coming from a boat.

Water waves seem complex, but particle waves share the same strange properties. Their broad structure and different velocities give rise to the uncertainty principle of Werner Heisenberg. Many people think this principle is original to quantum physics, but it is not; it was well known in the theory of waves and optics, developed in the 1800s long before it was proposed to apply to quantum physics.

Heisenberg made a precise statement of the uncertainty. A very short, narrow wave will have an accurate location, but such waves (whether for water or for matter) have a range of velocities; for many kinds of waves, the front of the wave group moves at a different velocity than the rear. Measure the velocity, typically by measuring the momentum (mass times velocity) and you may get any one of a large number of values. Measure its position and you'll get any value within the width of the wave. Virtually all waves will have some uncertainty in both their velocity and their position.

The math of the Heisenberg uncertainty principle follows exactly the math of classical waves. Appendix 5, "The Math of Uncertainty," shows this explicitly. The mathematical statement of Heisenberg's principle, often written as $\Delta x \Delta p \geq h/4\pi$,* is identical (except for the multiplication by Planck's constant h) to the equation that describes classical waves, including water waves, sound waves, and radio waves.

Uncertainty means that physics can no longer make definite predictions. It means that the future location of a particle cannot be predicted with precision, since to do so requires precise values of both the position and the velocity. Even worse, when combined with our current understanding of chaos, the small uncertainties engendered by quantum physics grow rapidly with time and deeply affect our macroscopic world. According to some theories, quantum uncertainty in the very early Big Bang is responsible for the existence of galaxies and clusters of galaxies, for the structure seen in Figure 13.4 (page 148).

Einstein did not like the uncertainty aspect of the new quantum physics, even though he had played a major role in inventing the field. Uncertainty implied that physics was incomplete, that somehow the future is determined by something other than the past. Quantum physics could not say what that was, only that it seemed to be random. In a 1926 letter to Max Born, Einstein wrote,

> Quantum mechanics is certainly imposing. But an inner voice tells me that it is not yet the real thing. The theory says a lot, but does not really bring us any closer to the secret of the "old one." I, at any rate, am convinced that *He* does not throw dice.

Werner Heisenberg recalled that at a conference, after Einstein made a similar remark, Niels Bohr responded, "But still, it cannot be for us to tell God, how he is to run the world."

---

* Δx is the uncertainty in the position; Δp is the uncertainty in the momentum; the symbol $\geq$ means "greater than or equal to"; h is Planck's constant.

The Littlest Distance

There is a very short distance that is possibly the smallest we can discuss with any meaning. (And it is not clear that we can even do that.) It is called the *Planck length*, and it comes from attempts to combine relativity theory with quantum physics. The Planck length is about 1.6×10^{-35} meter.

The Planck length is a consequence of the uncertainty principle, which implies that a small region of "empty" space cannot have zero energy, for if it did, the energy would indeed be certain. So, quantum physics typically ascribes a slight vacuum energy even to otherwise empty space. The smaller the region, the larger is this vacuum energy. If the region is sufficiently tiny, then the combination of large energy within a small radius will meet the requirements of the Schwarzschild formula, and the vacuum will have a microscopic black hole.*

Thus, combining quantum physics with general relativity seems to imply that the vacuum is microscopic foam of tiny but pervasive black holes. Moreover, each black hole rapidly fluctuates (appears and disappears) on a timescale given by the *Planck time*, the time it takes light to traverse one Planck length. Some theorists have proposed that perhaps space is digitized, like our computers, and that it exists only as discrete points separated by roughly a Planck length.

I have a generally broad criticism of all such speculation. The problem is that theory is overreaching experiment. Past theories were driven by measurement and experimental discoveries. If something happens, then it must be possible. There is no equivalent theorem for

---

* An estimate for the Planck length can be derived from the general principles stated in the text as follows: A typical minimum energy in a box of dimension L is given by quantum physics as $E = hc/(2\pi L)$, where h is Planck's constant. From quantum physics, the "zero point" energy is half of this, $hc/(4\pi L)$. For a black hole, set L equal to the Schwarzschild radius for mass $M = E/c^2$. That equation is $R_s = L = 2GM/c^2$. Combining these equations, we get $L = \sqrt{(Gh/(2\pi c^3))}$, the equation for the Planck length.

theory; if theory suggests it, it may or may not be true. These new theories, including all discussions of the Planck length, account for no experimental facts; they are driven by a desire for mathematical elegance. If that is the way to proceed, it is unprecedented in physics. We have essentially no tests of strong-gravity general relativity (it has been tested only in the weak limit, far from the regime of black holes), we have no compelling evidence for the properties of black holes (we only know that there exist massive objects that emit no visible light), and we have no experimental verification for black-hole radiation or black-hole entropy.

All the theory written on these subjects may be nothing more than fanciful speculation. That is not how physics has developed in the past. There may be many additional forces beyond the traditional "four forces" (electromagnetism, the nuclear force, the force of radioactivity also known as the "weak force," and gravity), and they may have to be discovered first so that they can be included in a proper theory.

Einstein fell into the trap, when developing his unified field theory, of trying to unify the wrong forces. Current grand unified theories may be making the same mistake.

Some theorists argue that there are no other forces, and they may be right, but I don't find their reasoning compelling. Gravity is an extremely weak force, and we never would have been aware of it except for two reasons: first, it has only one sign of its charge (all mass is positive), so it doesn't cancel itself; second, it has a long range, and can therefore be felt at very large distances as the force of many particles adds up. Any similarly weak force with canceling charges (as in electromagnetism, with protons and electrons) or short range would not yet have been discovered.

The uncertainty of quantum physics is amplified, in the world we sense and experience, by the phenomenon known as *chaos*.

The Uncertainty of Chaos

Versions of the following popular ditty date back at least as far as 1390:

For want of a nail the shoe was lost.
For want of a shoe the horse was lost.
For want of a horse the rider was lost.
For want of a rider the message was lost.
For want of a message the battle was lost.
For want of a battle the kingdom was lost.
*All for the want of a nail.**

These words illustrate the essence of modern chaos theory—that tiny causes can eventually lead to immense effects. In *Jurassic Park*, the pretentious mathematician Ian Malcolm describes the classic example of the *butterfly effect*: that (he says) a butterfly flaps its wings and as a result in Central Park, a week later, there's rain instead of sunshine. Popular use of the term *butterfly effect* predates chaos theory, going back to at least 1941, when it was described in G. R. Stewart's best-selling book *Storm*.

Chaos is observed in the motions of planets, in weather patterns, in population dynamics. The mathematical theory of chaos shows that the consequences of a small change can grow exponentially with time, at least initially. Therefore, infinitesimal precision would be needed to predict the future. The result is that, although we can predict weather a few hours, and sometimes even a few days, ahead, we are very poor at predicting it a week or a month ahead.

The effects of chaos are often bounded, however; sometimes the results simply switch back and forth between two very limited behaviors. The exponential doesn't continue forever. No matter how many

---

* A modernized version might start with "For want of a AAA battery, the computer mouse failed . . ." and end with thermonuclear war.

butterflies flap their wings, summer follows spring. Changes in climate require bigger forces than butterflies, such as changes in the orbit of the Earth, or the injection of billions of tons of carbon dioxide into the atmosphere. And despite Ian Malcolm's sententious pronouncements in *Jurassic Park*, we have no idea whether the flap of a butterfly's wing can change the behavior of a thunderstorm. His claim is speculation, not science.

Chaos theory does not deny causality or determinism. It simply means that extraordinary precision would be needed in measurements to be able to know what is going to happen in the long run. In that way, chaos is fundamentally different from Heisenberg's uncertainty. In quantum physics, precise values for both position and velocity are, in principle, unknowable. In a deep sense, these numbers don't even exist until measured.

When we combine chaos theory with quantum uncertainty, we reach the conclusion that tiny quantum uncertainty can influence even macroscopic behavior. Perhaps my own free will is determined by some quantum variation in a few atoms that works its way up in a chaotic chain into my nervous system and triggers behavior that, to all my friends and family and even to me, is unexpected and inexplicable.

Unfortunately, chaos theory is often grossly exaggerated in the entertainment world. In real physical systems, chaos usually operates within fairly narrow limits. The orbit of the Earth is chaotic, but the variations are quite small; they don't grow extraordinarily large, at least not over billions of years. We continue to travel around the sun in an orbit that is very close to circular (within a few percentage points). It has never been determined whether the flap of a butterfly's wing can actually trigger something large or whether its chaotic effects will remain small and local.

The movie *Jurassic Park* is full of exaggerations and misinterpretations of chaos. (The novel is somewhat more reasonable.) Malcolm warns, "You see a Tyrannosaur doesn't follow a set pattern or park schedules; [the dinosaur's behavior is] the essence of chaos." He pontificates that dinosaurs could never be contained, and he claims that this conclusion is a consequence of chaos theory.

That statement is utter nonsense. The best counter to Malcolm's exaggerations is one made by the movie's science adviser, paleontologist Jack Horner. Horner points out that the problems with the uncontrollable dinosaurs in the movie are all caused not by the inevitability of exponentially chaotic behavior, but by bad zookeeping. Lions and tigers and bears rarely escape from zoos; it was never inevitable that dinosaurs would. It all could have been prevented if the movie's character John Hammond, who built the park, had hired a zoo consultant.*

The Skeleton in the Quantum Closet

Nothing is more embarrassing to physicists than our complete inability to define what we mean by measurement. We chuckle when we teach about Schrödinger's cat, but deep down we know it is not a laughing matter. When we have no good response to a student's questions about the cat, we excuse ourselves; we are only following Feynman's advice to avoid thinking about it, lest we go down the drain.

There are books, chapters of books, conferences, and essays on the "theory of measurement." Google gives 239 million hits; Bing, 17.8 million. These results might give the misimpression that there is such a theory. If you look hard, you'll discover that what we really have is a collection of thoughts, many of which disagree with each other, and none of which have yet led to a satisfying conclusion.

One possibility is that for a valid measurement, a human must be involved—a sentient, self-aware, thinking soul. That idea was the very one that Schrödinger was attacking by putting a cat in the picture. Can you really believe that the cat is both dead and alive until a human peers

---

* *Jurassic Park* also depicts all vegetarian dinosaurs as kind and gentle and safe. I wonder whether screenwriter Michael Crichton and director Steven Spielberg believed the same for all vegetarian mammals, such as elephants, rhinos, buffalo, and hippos.

into the box? Martin Rees parodied this remarkable belief that no measurement is accomplished until a human is involved. He said,

> In the beginning there were only probabilities. The universe could only come into existence if someone observed it. It does not matter that the observers turned up several billion years later. The universe exists because we are aware of it.

I take this to be Rees's caricature of the egotistic notion that measurement requires human involvement—the same notion that Einstein mocked with his statement about the moon not existing until we look at it, the same one that Schrödinger tried to ridicule with his cat story.

Roger Penrose suggests that the universe itself makes measurements. We don't normally notice them, because they don't happen instantly; they take some time. The moon didn't need Einstein to look at it; it is far enough away that the universe somehow made it real before Einstein peeked. Penrose calls this "objective reduction" or "objective collapse." He speculates that this occurs "whenever two space-time geometries, and therefore gravitational effects, differ significantly." I sense that Penrose is on the right track, but his theory needs to be quantified; it needs to make predictions. Something causes the wave function to collapse long before it reaches humans. I don't know what it is, and I don't know how much time it would take, and Penrose doesn't claim he knows either; he is on its trail. Great thought is valuable, but we need to approach tough physics issues with experiments. Experiments with entangled variables (discussed in the next chapter) suggest that the magical time period is a minimum of a millionth of a second, at least in the laboratory.

Another attempt to deal with the measurement conundrum is called the many-worlds interpretation. I'll discuss it, too, in the next chapter.

So far, we have had one great experimental breakthrough—one that brings far more enlightenment to the issue than any gaggle of quarreling theorists. Stuart Freedman and John Clauser published their discovery in 1972. Their work proved that Einstein was wrong.

19

Einstein Is Spooked

*Einstein's contention that quantum physics is wrong is itself
proved wrong by a key experiment . . .*

> Everything we call real is made of things that
> cannot be regarded as real.
> — *Niels Bohr, a founding father*
> *of quantum physics*

> There are more things in heaven and earth . . .
> Than are dreamt of in your philosophy.
> — *Hamlet*

Einstein had the right word: *spooky*. He was talking about quantum physics, and there was something he thought was impossible. Ordinary quantum physics seemed to require wave functions to change faster than light. That couldn't be true. But it turned out it was true. Experiments show that it really happens, and if it happens, it must be possible.

The decisive experiment was done by Stuart Freedman and John Clauser at UC Berkeley. I remember being in awe of their exceedingly difficult project. They had to be extremely careful, since whatever result they found could shatter a whole class of theories and affront a whole class of theorists. Stuart became a close friend, and he used to joke about how he never discovered anything; he just proved that other physicists were *wrong*. Well, in this case the guy whom he proved wrong was Einstein, and I considered that to be a pretty big achievement.

One of Einstein's objections to quantum physics was the distressing feature of instantaneous wave function collapse. He called this collapse and other sudden changes "spooky action at a distance." A measurement of the position of a particle could, according to the Copenhagen interpretation, immediately, instantaneously, affect the amplitude of a particle that was light-years away. Einstein had previously shown in his theory of relativity that the very concept of instantaneous is meaningless for separated objects. Even the order in which events occur can depend on the reference frame. That meant that if one event caused another, in a different reference frame the causing event would come after the caused event (as in my tachyon murder paradox). Einstein explored this problem in a seminal paper with coauthors Boris Podolsky and Nathan Rosen, and their analysis became known (after their initials) as the *EPR paradox*.

Ah, but there was an easy solution, one favored by Einstein himself. He proposed a different interpretation of the wave function. It is not a physical object representing all of reality, but only a statistical function that reflects our uncertain knowledge. Einstein believed that the electron always has a real but hidden location, and that quantum physics simply didn't know what it was. No real wave disappears; no collapse has to take place. A hidden variable (such as the actual position) is missing in quantum physics. Add it in and physics again becomes *complete*, and once again the past completely determines the future.

There's an analogy in our understanding of gases. We don't know where every molecule is, but we have a theory that describes properties on average. The pressure we detect, and the temperature, are just averages of the properties of an enormous number of molecules. It is a statistical theory. The *ideal gas law*, relating the pressure of a gas to its volume and temperature, was just such a statistical average. Instantaneous pressure, measured just as a large group of molecule bangs into a wall, can be different, as seen in Brownian motion. Likewise for quantum physics, thought Einstein. He believed that hidden variables were the true theory and quantum physics was simply a statistical summary.

Then John Bell put teeth into the Einstein-Podolsky-Rosen paradox.

He proved that hidden-variable theories could not reproduce all of the predictions of quantum physics. That meant that quantum theory and hidden-variable theory were both falsifiable. One could determine which was correct by doing a suitable experiment. He analyzed a situation in which two particles were emitted in opposite directions (a setup that had been suggested by David Bohm) and argued that a good experimenter could determine which approach was right—the Copenhagen interpretation or hidden variables—by checking a limit that is now called *Bell's inequality*. Bell's work inspired John Clauser to search for an experiment that could demonstrate to the Copenhagen crowd that hidden-variable theory, Einstein's explanation for quantum behavior, was the right one.

The Hidden-Variable Killer

John Clauser was a young theoretical physicist who had just been hired at Berkeley by Charles Townes, the inventor of the laser. Clauser told Townes that he wanted to demonstrate experimentally that hidden-variable theory best accounted for physics results, and that the Copenhagen interpretation was wrong. Townes consulted with Eugene Commins, a professor who had developed experimental methods to observe what we now call *entanglement*. Commins and Townes agreed to jointly support the experiment. Commins's graduate student Stuart Freedman would do most of the experimental work.

Freedman and Clauser planned to search for the hidden-variable effect in photon emissions from a beam of calcium atoms—a choice suggested by their colleague Eyvind Wichmann, a great theoretician who (to my eye) always seemed to be bemused by controversy. They would measure the polarization—that is, the orientation of the two photons emitted from the calcium atom. Those photons should be similar, but the similarity predicted by quantum theory is different from that predicted by hidden-variable theory. I'll show this in more detail in a moment.

I knew both Freedman and Clauser (during this period I was first a graduate student at Berkeley and then a postdoctoral fellow), and I regarded the project as awesomely difficult. I'll hide that awesome difficulty here by assuming that both photons emitted by the calcium have identical, not just similar, polarizations. I'll assume that all the photons originate from the same spot, and that the atoms are not moving (not true in the actual experiment). I'll assume that the two photons are emitted in precisely opposite directions. I'll assume that the optical design is easy, not beset by aberrations. I'll assume that stimulating the atoms to emit photons introduces no other light that could confuse the detectors, and that there are no spurious reflections. I'll assume that detectors register photons with 100 percent efficiency, rather than the actual value of 20 percent. These simplifications will allow me to portray accurately the essence of the experiment, while deceptively making the experiment seem easier than it was.

Figure 19.1. Stuart Freedman, with the experiment that proved Einstein wrong.

According to hidden-variable theory (and with my simplifications), the two photons that are emitted by the calcium will come out in opposite directions but with equal but unknown polarizations. *Polarization* refers to the orientation of the photon's electric field; it is perpendicular to the direction of motion of the photon, but it could be vertical, horizontal, or anywhere in between. Many sunglasses have filters that cut out horizontally polarized light, such as the light that reflects off surfaces and makes "glare." If you tilt your sunglasses 90 degrees, then (with good glasses) all of the horizontally polarized glare will get through; you'll see lots of glare. If you tilt them 45 degrees, half will get through. Polarizing glasses have also been used for 3D movies, to let one eye see only horizontal light, and the other eye, only vertical. When two pictures are projected in the two polarizations, each eye sees a different image, giving the 3D effect.* Such glasses wouldn't work well outside the theater; glare is reduced for only one eye.

Back to the Freedman-Clauser experiment. Imagine now that the two photons come from the calcium atom in opposite directions. You place detectors on both sides with polarizers in front of each. You orient the polarizers perpendicular to each other, as shown in Figure 19.2. If both photons are vertical, then only the front polarizer will pass the light, and only the front detector will register. If both photons are horizontal, then only the rear detector will register. If both photons come out tilted at 45 degrees, then there is a 50 percent chance for each detector. That means that for such tilted photons, there is a 25 percent chance that *both* detectors will register a photon at the same time.

Amazingly, that is the prediction of hidden-variable theory, but not the prediction of quantum physics. In quantum physics, the tilted photon contains two amplitudes, one for vertical and one for horizontal. Those two amplitudes are like the amplitudes for a dead cat and a live

---

* Real 3D glasses typically use light polarized at +45 and −45 degrees; or they use "circular" polarization, which makes the effect insensitive to the angle of the viewing human head.

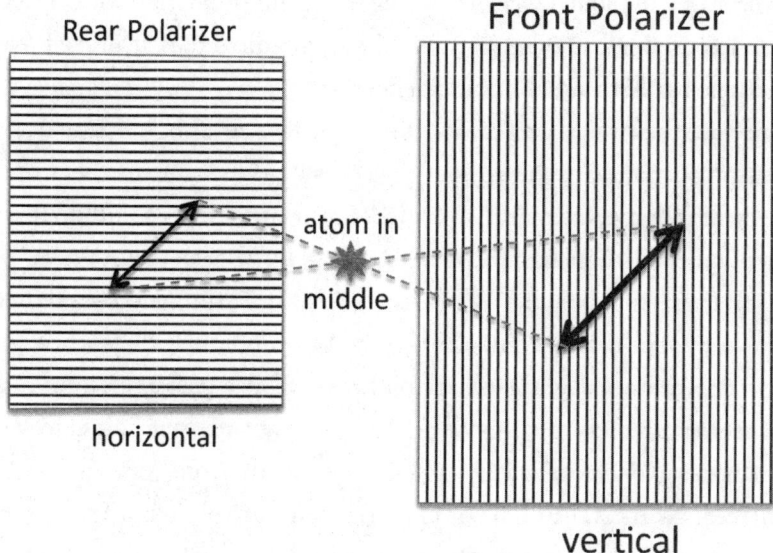

Figure 19.2. A simplified version of the experiment that proved Einstein wrong. Light polarized at a 45-degree angle has a 50 percent chance of passing each of the polarizers, but in hidden-variable theory, the probability for passage through the front polarizer is not correlated with the probability for the rear one. For quantum physics, passage through one implies no passage through the other, since the two photons are "entangled."

cat; the situation is not a mixture halfway in between, but a superposition of two possibilities. When one photon hits a polarizer—say, the vertical one—and passes through and is detected, then the amplitude of the other photon immediately changes. The horizontal component of the wave function disappears—collapses—leaving only the vertical component. Since the other detector is horizontal, the photon will *not* pass through.

No matter what the angle, once one photon is detected, the wave function will immediately collapse and the second photon polarization will never match the other, perpendicular polarizer. The result is the same regardless of the angle of polarization. The conclusion is that you

will *never* get a coincidence! That's the quantum prediction for this ide-
alized experiment. Hidden-variable theory predicts that, averaged over
all angles, 12.5% should be coincidences.

Suppose the two polarizers are miles apart—a million miles apart. In
quantum theory, as soon as one photon is detected, one of the amplitudes will collapse, disappear, everywhere, instantly, even a million miles
away. That's what Einstein referred to as spooky action at a distance.

Moreover, if both polarizers are vertical, quantum theory predicts
that every event will be coincident. Half of the photons will get through,
but whenever one does, the other polarizer will also pass a photon. Classical theory predicts that many photons will not register coincidences;
for example, if the polarization is at 45 degrees, then only one-quarter of
such events will get through both polarizers to register in both detectors.

Freedman and Clauser reported their results in 1972. Quantum theory, the Copenhagen interpretation, correctly predicted the experimental results. Hidden-variable theory was falsified. It was almost enough
to make one believe in ghosts. Unfortunately, Einstein had died in
1955. Spooky action at a distance had been observed in the laboratory.
Convincingly.

Clauser was dejected. According to Bruce Rosenblum and Fred Kuttner (in their book *Quantum Enigma*), Clauser said, "My own . . . vain
hopes of overthrowing quantum mechanics were shattered by the data."

Freedman and Clauser had shown that Einstein was wrong. There
are very few people in this world who have done that. Their work was
continued and improved on by Alain Aspect, who addressed some
of the possible loopholes that quantum-hating skeptics had come up
with. Rosenblum and Kuttner state unequivocally that they felt the
work was worthy of a Nobel Prize. I agree. Freedman and Clauser
had experimentally tested the Copenhagen interpretation, a founding assumption of quantum physics, found it to be superior to the
hidden-variable approach, and, along with Commins, triggered the
modern fascination with entanglement. I suspect the reason their
experiment didn't attract more attention was only that most phys-

icists ignored the problem. They did their best to avoid thinking about it, lest they go down the drain.

Entanglement

The Freedman-Clauser experiment is the clearest example of what is now widely called *entanglement*. Two particles are detected, far apart from each other, but they share a common wave function. Put another way, their individual wave functions (if you like to think of it that way) are entangled. The particles, when detected, could be separated by a meter or by a hundred meters or by a hundred kilometers, but detecting one immediately affects the detection of the other. It is immediate "action at a distance"—a nonlocal behavior unlike anything seen in prior theories.

It was still true that electric and magnetic and gravitational fields could not change faster than the speed of light, in accordance with causality. However, the quantum action at a distance is hidden, in the wave function or whatever other ghostlike quantum feature is running behind the scenes unobserved. The action at a distance takes place immediately, even though Einstein had taught us that *immediately* cannot mean the same thing in all frames.

We don't need two particles to have quantum physics violate relativity. That happens also with the infinitely fast collapse of the wave function of a single electron when it is detected. But the term *entanglement* is usually reserved for cases in which the wave function contains two or more particles. I think that's because the two-particle case seems somehow more disturbing.

If Einstein had been alive when Freedman and Clauser published their results, I think their experiment would have convinced him that his love for hidden variables was misplaced, that the Copenhagen interpretation is correct. But he would have hated being convinced. He complained that the Copenhagen interpretation implied that quantum physics was incomplete. Perfect knowledge of the past could not give a perfect prediction of the future. There must be a better theory.

Later I'll argue that not only is quantum theory incomplete, but perhaps physics, and all science, is fundamentally incomplete.

Messages Faster Than Light

Could we use the collapse of the wave function to send instant signals over arbitrary distances? Could the Freedman-Clauser two-photon method be used to send faster-than-light information from one polarizer to the other? Many people, when thinking about this, believe that there must be a way. Perhaps I could send a signal by choosing to try to detect one photon, or by not trying to detect it. But if you think about it for a while, you'll realize you can't signal in this manner. At the distant detector half of the photons will still be observed. No information will be detectable by a person at that location. The detected photons will appear to be a random selection of the photons arriving. The distant experimenter will have no way to recognize that his measurement results correlate with yours.

Maybe I could send a message by changing the orientation of my polarizer? No; that doesn't work. The detections at the distant location will still appear random. They won't be truly random; they will be correlated to the photons that I detect, and those depend on the orientation of my polarizer, but they will still seem to be random. The attempt to send information fails because the experimenters have no control over when they detect the particle.

All attempts at figuring out how to use the collapsing wave function to send a signal faster than light have failed. Try to find one yourself— but don't spend too much time on it. We now know your effort will be fruitless. In 1989, the no-communication theorem* was proved showing that if the rules of quantum physics and the Copenhagen interpretation

---

* The no-communication theorem was originally proved in 1989 by Philippe Eberhard and Ron Ross, two of my colleagues at Berkeley, and then elaborated by others, particularly Asher Peres and Daniel Terno in 2003.

are correct, you can't send information using the collapse of the wave function—not at super-lightspeed, not at any speed.

I wonder whether this theorem would have pacified Einstein's opposition to quantum theory. It shows that no measurable quantity violates relativity; only the unmeasurable wave function does that. I suspect it would not have mollified him. Having any structure in a theory that violates relativity is a concern, even if it is an undetectable one. And it is still true that quantum theory is incomplete; it contains an element of a random God throwing dice that Einstein despaired was undermining physics.

Other work on measurement theory continues. In Chapter 21, I talk about the no-cloning theorem, which says you cannot duplicate an unknown wave function unless you destroy it. That prevents me from making thousands of copies of a wave function and then probing each one in a slightly different way to discern its detailed structure. That structure is beyond our ability to measure. It will remain forever ghostlike.

Crutches

Throughout my early career, I had a way of dealing with spooky action at a distance. I simply believed that the wave function was a crutch, something that had proved useful for thinking about quantum physics but wasn't really needed. Someday, a theory avoiding it would be created—a theory that had no collapsing wave function. But the Freedman-Clauser experiment threw a wrench into my hope. Detection at one polarizer affects detection at the other, even though the two events are not "connected" by lightspeed, even though they are so far apart that the answer to the question of which detection came first depends on reference frame. Spooky action at a distance is not merely a feature of the theory; it is a feature of reality.

There is a long history of crutches in physics—concepts that were introduced to help in the initial understanding and acceptance of a theory, but later abandoned as unnecessary and possibly misleading. In his electromagnetic theory, James Clerk Maxwell imagined that space was full of little mechanical gears that transmitted radio waves and light. The diagram in Figure 19.3 is from his original paper. It shows space full of little rollers and wheels that mechanically convey action across distances. Perhaps this is the way Maxwell actually imagined it, or perhaps it was only Maxwell's way to convey his concept of electromagnetism to other physicists who were comfortable with mechanics but not the new abstract concept of "fields" that spread out across otherwise empty space.

Now Maxwell's original diagrams are referenced for amusement only, to show students how even a great theorist can paint silly pictures. But if light is a wave, what is its medium? A new crutch was soon devised: "aether," the material that was waving when electromagnetic waves waved. The concept of aether was undermined in 1887 when Michel-

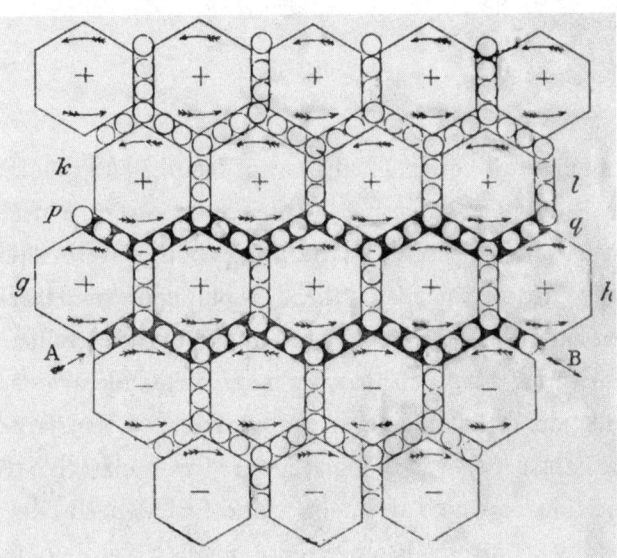

Figure 19.3. Maxwell's depiction of the way he thought about empty space— an array of mechanical wheels. This diagram was more physically plausible to nineteenth-century scientists than was the abstract concept of a "field."

son and Morley failed to detect the aether wind. Einstein showed in his relativity theory that such motion was undetectable, since the velocity of light is constant in all directions. In a sense, the aether was like the quantum wave function: unobservable.

Present quantum theory is still stuck with instantaneously collapsing wave functions. They do no good; they are undetectable; we can't use them to send signals. Some kind of cosmic censorship seems to separate them from true reality. (Recall that in Chapter 7 I spoke of black-hole censorship that allowed time "beyond infinity.") I expect that, someday, instantaneously collapsing wave functions will be found to be unnecessary for calculations, and they will be forgotten. But that day hasn't come yet, because we haven't yet figured out how to do calculations without them.*

But the Freedman-Clauser experiment suggests that there will be a causality issue, regardless of the use of wave functions. The results of one experiment can affect the results of a distant experiment with a speed exceeding that of light.

What's Wrong with "Spooky"?

Standard quantum physics—the Copenhagen interpretation—has spooky action at a distance. So what? None of its predictions for experiments violate relativity, so who cares? Well, for one, I do. I get only minor comfort from the fact that such collapse can't be used for super-lightspeed communication. The collapse of the wave function at infinite velocities bothers me, and I take that as a hint that the formulation is wrong. Many other physicists agree. That's why they keep attending conferences on the "foundations of physics." They smell something fishy, and I don't mean fresh fishy. For a potentially great discovery, they are willing to risk going down the drain.

---

* Note for the experts: The Heisenberg picture, an alternative formulation of quantum physics, does not have explicit wave functions, but it has state vectors, and they, too, change at infinite speed.

At one of these recent conferences, attendees were asked to vote for their favorite interpretation of quantum physics. Remarkably, a strong plurality of 42 percent picked the Copenhagen interpretation.* The next highest was an "information-based" interpretation, with 24 percent. An intriguing idea called the *many-worlds interpretation* was down at 18 percent. Less popular was an idea I mentioned in Chapter 18, Penrose's "objective collapse"—that the universe is constantly measuring itself— with only 9 percent. (That would have gotten my vote.)

It is fascinating that, even at a conference that attracts those thinking most deeply about these issues, the Copenhagen interpretation is far out in the lead. Despite its spooky character, it has held up to experimental tests.

Indeed, some of the alternatives are just as spooky. The many-worlds interpretation gets a lot of attention (despite its low ranking in the poll), perhaps because it has, by far, the most flamboyant name. It posits simply that wave functions never collapse; in the depiction of Schrödinger's cat in which the film strip divides (Figure 17.1, page 195), both futures take place. That figure shows two worlds, but in "reality" there will be an infinite number, as the film continues to split every zillionth (or smaller) of a second.

I consider this scenario just as spooky as the infinitely quick collapse of the wave function. Which of the many worlds, the infinity of universes, do I experience? Somehow my soul picks one. But someone else may be traveling along a totally different path—yet I am part of that universe too. I'd rather accept action at a distance than the picture that I exist in an infinite number of universes simultaneously.

Is that a failure of my own imagination? Perhaps, but the only potential value of the many-worlds picture would be to mollify my imagination; the theory itself is not testable. It makes no predictions that distinguish it from the Copenhagen interpretation. Nonetheless, some of its proponents, most prominently Sean Carroll, say it is self-evident.

---

* See "The Most Embarrassing Graph in Modern Physics," posted on Sean Carroll's blog January 17, 2013, http://www.preposterousuniverse.com/blog.

(That claim is both *ipso facto* and *ipse dixit*.) Supporters say it simply reflects the equations and avoids the need to confront the meaning of *measurement*. In doing so, it substitutes a new concept: the fact that each of us exists in many worlds but experiences only one. I don't know whether you find that spooky, but I do.

Computing with Ghosts

Back when Freedman and Clauser were doing their work, the field of quantum measurement was largely ignored. But recently it has become a hot field, with funding not only from the National Science Foundation and the Department of Energy but also from the Department of Defense, the CIA, and the NSA. The reason is the fantastic potential of quantum computing.

The essence of quantum computing is that you can store and manipulate information in wave functions. There is a vast advantage to avoiding using ordinary bits, with their limited ones and zeros, and instead using *qubits*, each consisting of a quantum amplitude. A qubit can be manipulated and used in a computation. In an important sense, a qubit contains far more information than does an ordinary bit. Consider, for example, the quantum wave function in the Freedman-Clauser experiment. The ratio of the two polarization amplitudes is analogous to the classical angle of polarization, with any value between 0 and 90 degrees. That's a lot more information than storing a 1 or a 0. That qubit is a superposition of two states, and the information lies in their ratio. The catch is that you can't retrieve that number. You can only get a probability of up-down or right-left polarization.

The fact that you can't measure the wave function but can only sample it (and in the process, making it collapse) doesn't mean you can't compute with it. Wave functions are affected by forces and interactions, so the wave function can be manipulated even without a measurement. For example, even though a polarization can be detected only probabilistically, the polarization wave function can be rotated exactly. The trick in quantum computing is to do all the manipulations on the invisible wave function, stored in qubits, and then make a measurement only when the computation is finished. That final answer may be one that can be stored in a few qubits, even if not all of the computation can be.

Imagine you have a very large number, perhaps with 2,048 digits, and you would like to factor it. (Factoring is key to breaking some advanced cryptography.) You don't care about all the attempts at factoring that fail; all you really need are the two roughly 1,024-digit numbers that are the factors. This is the hope of quantum computing; this is (partially) why intelligence agencies support the research and development. Quantum computing could enable enormously complex computations to be done in parallel. They could be done without generating heat. Every time a bit is flipped in an ordinary computer, a minimum of heat is generated.* But with quantum computing, you (in principle) generate heat only when you finally measure the qubit at the end.

Will quantum computing succeed? I am pessimistic. Some simple computations (factoring 6 into 2 × 3, and factoring 15 into 3 × 5) have already been achieved, but complex computations are much harder. In fact, not only am I pessimistic, but many people working hard in the field are secretly pessimistic too. Then why are they doing it? I think the reason is that they are fascinated by the issues of quantum measurement. Thanks to quantum computing, there is finally money available for studying what happens when you manipulate and measure quantum systems. Already we have wonderful new theorems, such as the no-communication theorem, that (in principle) could have been proved back in the 1940s. And if their work yields a breakthrough in our understanding of quantum measurement, it could lead to yet another revolution in physics.

---

* According to physics theory, the minimum heat generated is given by $\sqrt{2}kT$, where k is our old friend, Boltzmann's constant, from statistical physics, and T is the absolute temperature.

20

Backward Time Travel Observed

A positron—later postulated by Feynman to be an electron moving backward in time—is discovered . . .

Now, if my calculations are correct, when this baby hits 88 miles per hour . . . you're gonna see some serious shit!
— *Dr. Emmett Brown, triggering time travel in*
Back to the Future

On August 2, 1932, Carl Anderson discovered what appeared to be an electron with the wrong charge, positive rather than negative. In his paper, he referred to it as a "positron" and identified it as the antimatter that had been predicted by Paul Dirac just over a year earlier. Seventeen years later, Richard Feynman suggested that what Anderson had seen was an electron moving backward in time.

The photo in Figure 20.1 was taken by Anderson using a cloud chamber, a device that records rapidly moving electrons and protons by the cloud particles that condense along their paths, little droplets of liquid, the small dark dots in the photo. In the image, the positron enters from below, passes through a thin lead sheet, and then comes out on top. The path curves because Anderson had the cloud chamber in a strong magnetic field. The fact that the path curved to the left told him it had a positive charge, like that of the proton, but the *way* it curved told him it was much, much lighter. The fact that the curvature was greater on the top

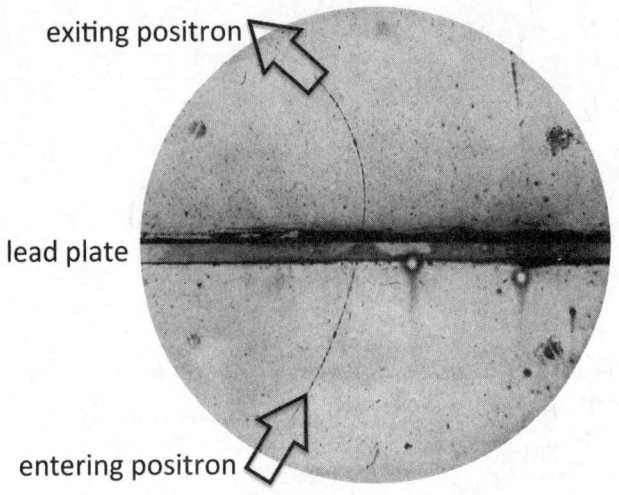

exiting positron

lead plate

entering positron

Figure 20.1. Anderson's photo of the first identified positron (antielectron); notations added. After passing through the lead plate, the positron curves more in the applied magnetic field because it has slowed. Or is this an electron moving back in time?

part of the image told him that the particle had slowed, confirming that it had entered from the bottom.

It may seem strange to describe this as an ordinary electron moving backward in time, but that has become the standard way to treat these particles in advanced quantum calculations, an approach devised by Richard Feynman. Backward time motion has become one of the standard tools of physics, one that many physicists use every day. Graduate students are taught how to apply these reversed time methods in advanced quantum classes. Even "simple" calculations, such as the bouncing of one electron off another, involves particles (typically photons) moving backward in time.

No one would introduce backward time travel without compelling reason. In this case, one compelling reason was Dirac's ridiculous theory of the positron that had preceded Feynman's work.

The Most Absurd Theory in This Book

At the time Anderson saw his positron, he did not think it was an electron moving backward in time. He thought it was a bubble, an emptiness, a moving hole in an infinite sea of negative-energy electrons that densely filled all of space. I am serious. As absurd as that sounds, it was the basis of the prediction that Anderson was confirming. It wasn't Anderson's idea, but the concept of Paul Dirac, the man who had managed to unify the new quantum ideas (of an electron being a wave) with Einstein's theory of relativity (although he did not address the issue of the instantaneous wave function collapse).

Schrödinger's equation was nonrelativistic; it didn't include any of the effects that Einstein had in his theory of relativity. Dirac had approached the challenge of creating a relativistic quantum theory of the electron. He took what he considered to be a logical and straightforward

Figure 20.2. Paul Dirac, the father of antimatter.

approach. He made a judgment about what the equation should look like (in particular, he decided it had to have a simple time dependence) and then worked out the math. The math turned out to be remarkably complex, challenging for even an advanced graduate student in physics to understand fully. But it satisfied Dirac's goal of keeping the time dependence simple.

The equation that Dirac found worked amazingly well. With no need to introduce adjustable parameters, it automatically included the previously known fact that the electron has spin, correctly gave the allowed values of that spin, and even accounted for the fact that every electron was not only a small electric charge but that it was also a small magnet. With a simple assumption, Dirac's equation gave a precise and accurate value for the strength of that magnetism.* He published his theory in January 1928. The equation was a remarkable success. It was perhaps the most outstanding work in theoretical physics since Einstein had correctly accounted for the advance of the elliptical orbit of Mercury using general relativity.

There was one little (actually, huge) problem. Dirac's theory predicted that the electron could have either positive rest energy $+mc^2$ or negative rest energy $-mc^2$. That was pretty bad; nobody had ever observed negative mass. But perhaps even worse, the existence of negative-energy states implied that electrons were unstable. A positive-energy electron would spontaneously jump from its positive-energy state down to the negative-energy state, losing $2mc^2$ in the process (presumably by radiating photons). No positive-mass electron would last for even a millionth of a second before turning into a negative-mass electron. Yet positive-mass electrons are the known

---

* Note for the experts: Dirac did have to make an additional assumption—"minimal electromagnetic coupling"—in order to obtain the magnetic moment. That was the only additional assumption he had to make. That assumption turned out to be false for other particles thought to be elementary at that time, such as the proton. We now know the proton is a composite made up of three quarks and a few other things.

ones; they don't decay. No negative-mass particles had ever been seen. In his first paper, Dirac explicitly stated he was ignoring that problem for the time being, but because of it, he considered the theory not yet final. He wrote,

> The resulting theory is therefore still only an approximation, but it appears to be good enough to account for [the known spin and magnetism of the electron] without arbitrary assumptions.

Dirac "solved" this negative-energy problem two years later with one of the most extraordinary (I would call it ridiculous) proposals ever made in physics. It was known that atoms could hold only a limited number of electrons. That's because their *orbitals*, the possible locations the electrons occupied, could take only two electrons each. (That ad hoc rule had been proclaimed by Wolfgang Pauli, and is now called the *Pauli exclusion principle*. It was later put on a firmer foundation as the theory of quantum physics was developed.) So, Dirac proposed a similar solution for empty space. He suggested that all the negative-energy states, an infinite number of them, were already filled with negative-energy electrons. The vacuum was so chock-full of electrons with negative energy that there was no room for any more. Positive-energy electrons could not lose energy and fall into one of these negative-energy orbitals because those orbitals were already all filled up. He referred to empty space as a *sea* of negative-energy electrons, filled to the brim.

Wouldn't this imply that all of empty space is not empty, but has infinite charge and also infinite (but negative) energy density? Yes. How could that be? Wouldn't we have noticed? Dirac said no. That's what the vacuum is. Because the charge is uniformly spread, we live within it without being aware of it. Do fish notice water? All of our physics is based on what happens in the midst of this constant background. We are not aware of this infinite sea of charged particles because it never changes. Dirac's vision made Maxwell's picture of tiny rotating wheels look simple.

Dirac's huge negative-mass density would be expected to have huge gravitational effects, but he never addressed that issue, presumably because the expansion of the universe discovered by Hubble had been announced only nine months earlier, and the understanding of the dynamics of that expansion, published in an obscure journal by Lemaître, was not yet widely known. The gravity effects of Dirac's negative sea are related to the modern problem that a theoretical calculation of the dark energy is incorrect by a factor of 10^{120}, as mentioned in Chapter 14.

A lesser physicist would have proclaimed a new "exclusion principle" postulating that negative-energy states are excluded; electrons can't occupy them. Not Dirac. He said that if the equation had such states, they must exist, and the problems caused by their existence must be dealt with. The best solution he could find was the infinite negative-energy sea. He never attempted to explain how this infinite sea had been created, or why it was filled only from negative infinity to 0, and why there was no positive-energy sea of filled states.

Only in a physics world that had been recently hit by so many mind-bending surprises (time dilation, length contraction, curved space-time, quantized packets of light) would such a ridiculous proposal be taken seriously. It was. Maybe it was not absurd, but brilliant. In fact, even today it lends psychological support to those who make modern outlandish claims, such as the idea that we live in an eleven-dimensional space-time.

Dirac took his idea even further. From time to time, one of the negative-energy electrons in this infinite sea might be hit with another particle, gain energy, and leave the sea. It could jump to a positive-energy state (those were not filled). What would be left behind would be a bubble, what Dirac called a hole. That hole could move through the infinite sea just as bubbles move through water (it is primarily the water outside the bubble that is moving, not the small amount of gas inside it), and the absence of a negative charge in the sea of negative charge would behave as if it were a positive charge.

Was that a prediction of antimatter? Not yet. Dirac proclaimed that such holes were protons! He wrote an article in December 1929, "Theory of Electrons and Protons," laying out this concept.

Dirac Reluctantly Predicts Antimatter

There was a serious problem with Dirac's bubble proton theory. Hermann Weyl showed that the bubble would move as if it had the same mass as that of the electron—but the proton was known to be 1,836 times heavier. Being wrong by a factor of 1,836 is not completely unprecedented, but it was a challenge. Dirac had no good answer for the mass discrepancy. The theory was still tentative; it would have to be improved further. He referred to a recent calculation by Eddington as offering some hope, but the proton mass discrepancy was a serious unsolved problem that had to be addressed.

Another severe problem emerged three months after Dirac's proton paper appeared. Robert Oppenheimer, later to gain fame as the leader of the Manhattan atom bomb project, wrote a paper pointing out that Dirac's protons, his holes, would be attracted to electrons, and when they met, the pair would *annihilate*, destroy each other and emit all their mass energy as gamma rays. Neither protons nor electrons could survive in ordinary matter for more than a millionth of a second. But that doesn't happen; electrons and protons live happily together in atoms; they don't annihilate. Dirac's theory contradicted the most basic experimental observation.

Weyl and Oppenheimer were right. Dirac's theory was in deep trouble. But how could his theory be wrong when it correctly explained the spin and magnetism of electrons?

Finally, in May 1931, Dirac wrote a paper that mentioned a desperate solution. Remarkably, most of this paper was about a totally different topic, the connection between electric and magnetic fields. Its title, "Quantised Singularities in the Electromagnetic Field," gives no hint that it includes a brief comment on the negative-energy problem; of thirty-six paragraphs in the article, only two touch on it. Dirac seems reluctant about the solution he has been forced to invent—the prediction of antimatter. In his paper, he says,

> A hole, if there were one, would be a new kind of particle, unknown to experimental physics, having the same mass and opposite charge to an electron. We may call such a particle an anti-electron.

Dirac explained that anti-electrons are not present in nature because they quickly annihilate with electrons, just as Oppenheimer said they would. That's why we don't see them. In principle, anti-electrons could be created in a laboratory using high-energy gamma rays, but Dirac thought that doing so was beyond the technology available at that time. He wrote,

> This probability is negligible, however, with the intensities of gamma-rays at present available.

It is much more comforting to discover that you can explain a previously known mystery, such as the magnetism of the electron, than to be forced to make a prediction. If antimatter exists, why hadn't anti-electrons been seen? Dirac was not an experimentalist, and he had limited understanding of the true limits and capabilities of the then current experiments. With greater knowledge, he might have been even more concerned about his prediction, because experimentalists for several years had been capable of observing his predicted anti-electron. His defensive excuse that the "probability is negligible" was completely wrong.

We now know that Dirac's anti-electrons had already been observed—not produced from laboratory gamma rays (Dirac was right on that), but created by the high-energy cosmic rays. Cosmic rays are a natural radiation observable on the Earth's surface that originate from space—a fact demonstrated by physicist Victor Hess in the 1910s. Primordial cosmic rays colliding in the atmosphere produce anti-electrons and other antimatter. In 1927, a year before Dirac's original electron theory was published, and over three years before Dirac predicted the positron, Russian scientist Dmitri Skobeltsyn had likely seen positrons in his cosmic-ray studies. However, he had no way to measure charge (positive versus negative), and he had no way to see annihilation, so he couldn't distinguish matter from antimatter.

In 1929, also prior to Dirac's anti-electron prediction, Chung-Yao Chao, a physicist at Caltech working in a room close to Carl Anderson's, had seen a strange effect in the way cosmic-ray electrons (at least that's what he thought they were) were absorbed in matter. Their behavior didn't match expectations. After Dirac's theory came out, Anderson correctly guessed that the difference could be accounted for if anti-electrons were mixed in. That interpretation inspired him to set up an exquisite cloud chamber with a strong magnetic field and a lead barrier that could identify which way the particle was moving (since it would lose energy in passing through).

Anderson made his discovery and published the photo shown in Figure 20.1 (page 232) and convinced everyone that antimatter existed. Dirac was right. The editors of the journal suggested to Anderson that he name the particles *positrons*, and the name stuck.

My mentor Luis Alvarez knew Anderson and greatly admired his work. He told me about one of Anderson's concerns that I don't think has been previously reported. The 1930s were an era when the practical joke was in high fashion among college students. Alvarez himself was proud of some pretty clever tricks he had pulled on other physicists, particularly on arrogant professors. So Anderson, armed with his first photo of an anti-electron, worried that someone had played a trick on him. All a jokester had to do was insert an extra mirror in front of Anderson's automated camera, and an electron would have appeared to curve the opposite way. So, he carefully reexamined the photo and compared it with the apparatus, but he concluded that the photo was indeed legitimate. He published, and history was made.

The next year, 1933, Dirac accepted the Nobel Prize in Stockholm, Sweden, for what was then known as the "theory of electrons and positrons." His Nobel lecture explains what he did but makes no mention of Weyl or Oppenheimer or Anderson.

Aether Reborn

After Einstein and until Dirac, vacuum had been thought of as empty space. Einstein had shown that motion with respect to absolute space was undetectable, so it made no sense to talk about the makeup of nothing. Aether appeared to be dead, gone from the physics lexicon. The vacuum was absence of anything; like the number zero, it didn't exist. Then Dirac claimed it was loaded with negative-energy electrons. Not only was the vacuum made of something, but it had infinite negative charge and infinite negative energy.

Despite all this structure to the vacuum, motion through it still could not be measured. Dirac's theory had been constructed within the math of Einstein's relativity, and motion with respect to the brimming sea of negative-energy electrons was undetectable. In some sense, the old aether had been reborn. In fact, maybe that infinite sea provided the medium that waved when light propagated. Electromagnetic waves were analogous to ocean waves, but instead of moving through water, they moved through a different sea—the infinite sea of negative-energy electrons.

In my undergraduate courses on electromagnetism at Columbia, I was taught that aether didn't exist, that the aether concept had been proved unnecessary and irrelevant and had been abandoned. But then in graduate school at UC Berkeley, my professor Eyvind Wichmann (the same person who suggested using calcium in the Freedman-Clauser experiment) pointed out with a smile that aether never had disappeared from physics; it had simply been renamed. These days we call it the *vacuum*.

Look up *vacuum* in a graduate physics text. You'll find that the vacuum is far more complicated than was Maxwell's aether. It is *Lorentz invariant*, meaning that you can't detect it by the fact that you may be moving through it; there is no vacuum "wind." The vacuum contains energy. It can be polarized; that is, it responds to an electric field by separating its "virtual" charges. That polarization can be detected and measured by looking at the energy levels in the hydrogen atom (through something called the Lamb shift) and can be detected directly by the force the vacuum can exert on metal plates (the Casimir effect). We now think of the vacuum as constantly producing matter and antimatter, which almost as promptly annihilate—except when near a black hole. This feature grew to prominence in Stephen Hawking's theory of black-hole radiation; a heuristic explanation of the radiation has the intense gravitational field close to the Schwarzschild surface separating the background matter and antimatter pairs before they annihilate, sucking one into the black hole and emitting the other to infinity.

The modern view of vacuum treats it as a thing. It doesn't move (at least in a detectable way), but it can expand, and that fact is important for understanding the Big Bang. It contains a constant Higgs field, filling all of space, responsible for giving particles their mass. It also contains a dark energy that is responsible for acceleration of the expansion of the universe. It is far more complex than that bunch of gears and wheels that Maxwell imagined.

Feynman Reverses Time

It took seventeen years to vanquish Dirac's infinite sea. Maybe it would have happened sooner, but a nasty war intervened that distracted the hero of backward time, Richard Feynman. Feynman worked on the Manhattan Project, observed the first atomic bomb explosion, and returned to fundamental physics at Princeton where he lectured in front of the monster minds showing that radiation doesn't show time asymmetry. Feynman was a great polymath, someone who illuminated virtually every physics problem he thought about. And he thought about many aspects of physics, from electromagnetism to particle physics to superconductivity to statistical physics.

In Dirac's equations—indeed, in all of the equations of quantum physics—the energy term always appears in combination with time, as a product Et. Dirac's positrons contained that term with a minus sign: $-Et$. (This combination was a consequence of Emmy Noether's work, discussed in Chapter 3.) Dirac interpreted the minus sign as showing negative energy. Feynman suggested instead that the equations could be indicating positive energy combined with negative time. Time moving backward may sound ridiculous, but is it any more ridiculous than an infinite sea of negative-energy electrons?

Feynman had not been the first to consider backward time, but he was the first to turn it into a detailed theory. He suggested that a positron actually was an electron moving backward in time. That explained why it had the same mass as the electron; it *is* an electron, and it has positive energy. In fact, the electrons kept their negative charge; their motion backward in time would give the illusion of positive electric charge. The infinite negative-energy sea was gone; the negative sign had been transferred from energy to time.

Feynman had developed a whole new way to think about quantum physics, particularly the physics of *fields*—those lines of force that emanate from charges and magnets. Feynman discovered a series of equations that could be used to calculate all quantum processes in electromagnetism— and then he realized something even more fascinating. Each of his equations could be depicted as a simple diagram. Rather than working out complicated equations when you're given a new problem to calculate, try drawing every possible diagram you can imagine within the rules he created, and then, using another set of rules, write down the corresponding equations, and you will have your answer, the quantum physics amplitude that the process (typically a collision between particles) will take place. The result was so simple, so spectacular, that Feynman speculated that the diagrams might be more fundamental than their derivation.

Feynman's approach made quantum physics so intuitive that now most physicists think in terms of these *Feynman diagrams*. Suppose, for example, we want to know how an electron and a positron would behave if they bounced off each other in space. One Feynman diagram is shown in Figure 20.3.

This simple diagram could be called an "annihilation" diagram, since the positron and the electron vanish, turning into a photon, which then decays back into an electron and a positron. In the Feynman approach, this diagram corresponds to a specific equation, one that gives the amplitude for the scattering, and from that we can calculate the probability of the scattering.

But from Feynman's rules, based on the equations, you have to add in another amplitude, corresponding to the diagram shown in Figure 20.4. This picture can be called an "exchange" diagram. Just as in the previous diagram, the electron and positron enter from the left and leave on the right. But here the particles that emerge are the same ones that entered. The scattering comes about from the positron and the electron exchanging a photon. The photon exchange causes the electron and the positron

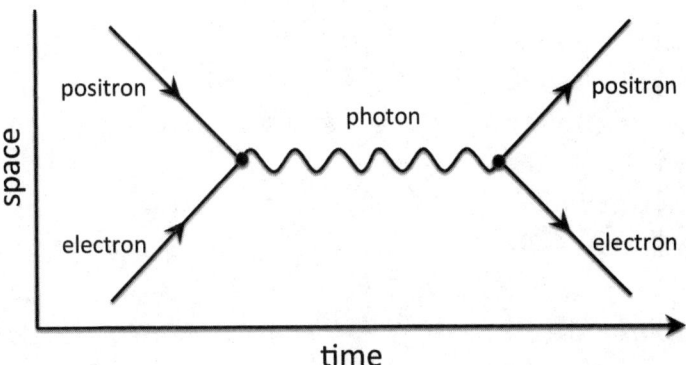

Figure 20.3. A Feynman diagram showing one way that an electron can scatter (bounce) off a positron. The positron combines with the electron to form a photon, and then the photon decays back into a positron and an electron.

to change their paths; it provides the equivalent of the force between them. Note that the whole concept of force has been eliminated; the electron is deflected not by a force, but by absorbing a photon. In both of these diagrams, the photon is hidden from observation; it is there only temporarily and is called a *virtual photon*. Because it is short-lived, it doesn't even have to be massless; in Feynman's theory, virtual photons typically do have rest mass.

To calculate the total amplitude, the quantity that gives you the total probability of scattering, you add together the amplitudes for each of the diagrams. That seems reasonable, until you think about it. In the first diagram, the original electron disappears and a newly produced one emerges on the right. In the second one, the same electron enters and leaves. Yet the two processes are happening simultaneously. Physics can't tell whether the emerged electron is the same one or not. In fact, it both is and isn't. The particles are truly identical, indistinguishable. To repeat, the emerging electron is both the same one that entered and a

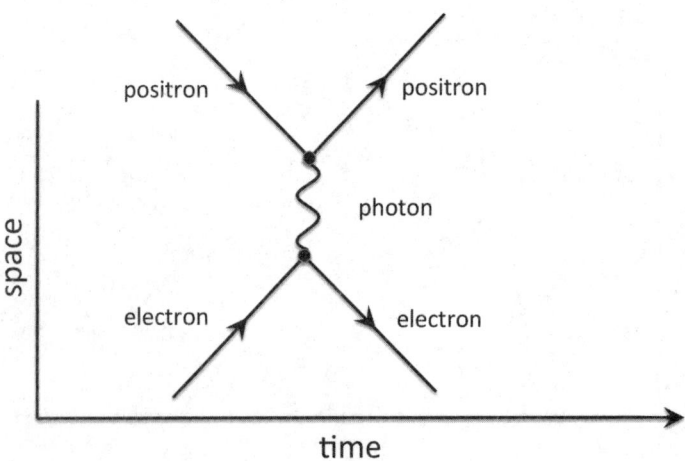

Figure 20.4. Another way that an electron and a positron can scatter off each other. In this Feynman diagram, the electron and the positron exchange a photon.

different one that was created, simultaneously. Shades of Schrödinger's cat! The probability for the process consists of the amplitudes for both diagrams added together, and then the combination is squared.

Remember Feynman's advice. Don't think about how this can be so or you will go down the drain.

Let's now return to backward time. In Feynman's new approach to positrons, the first diagram (Figure 20.3) is completely equivalent to (that is, it gives the same amplitude as) the diagram in Figure 20.5. Note the small changes. What was once seen as a positron is now an electron moving backward in time.

The Feynman diagrams are the key element in current quantum physics calculations, and every day thousands of people use them. There are computer programs to evaluate the amplitudes of complicated Feynman diagrams (such as those with two or more photons exchanged; for some examples, see the background of the Feynman stamp in Figure 16.1, on page 181). In these, antimatter is represented as ordinary matter moving backward in time. Moreover, when the particles move backward in time, they carry information with them about the future. They bring

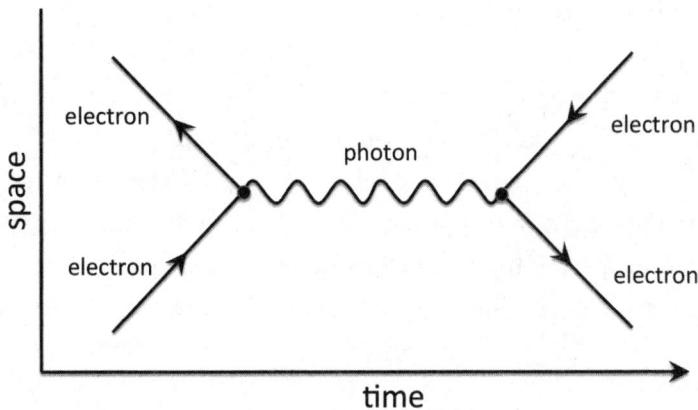

Figure 20.5. This is the same as Figure 20.3, except that the positrons are now depicted as electrons moving backward in time.

with them the momentum and energy of the future particles that appear on the right-hand side of the diagrams. Feynman says this approach was inspired by his work on radiation, the work he presented in front of Einstein and the other monster brains showing that classical radiation works both forward and backward in time.

Despite its disturbing features for our sense of reality, Feynman's backward time causes no problem with physics, because physics equations don't need or use the flow of time.

Hawking refers to Feynman's reverse-time paradigm in his *Brief History of Time* but is reluctant to accept it as backward time travel. He states (without explanation) that he believes such motion backward in time is possible only in the microscopic world, but not in the larger world of humans.

Is it possible that all electrons are really positrons moving backward in time? Or that we are made of positrons, and that the electrons in our bodies are positrons moving backward in time? Yes, all of these suggestions are not only possible, but part of current theory—or, as some people argue, one way to interpret the current theory.

Which way is right—Dirac's or Feynman's? Are positrons bubbles in an infinite sea, or electrons moving backward in time? Most physicists prefer the Feynman picture. It seems to satisfy Occam's razor, the principle that the simplest explanation of a phenomenon is the one we should accept. But there is no empirical way to demonstrate that backward time travel is right and the infinite sea of negative-energy particles is wrong. And it is certainly possible that neither view is correct. The Feynman diagrams were all derived from quantum field theory, and we may be overinterpreting them when we take them literally, rather than as mnemonics for remembering the Feynman equations. But maybe not.

We Are All One

In his book *Surely You're Joking, Mr. Feynman*, Feynman wrote that he once got an excited phone call from his mentor John Wheeler, who said,

"Feynman, I know why all electrons have the same charge and the same mass!" Feynman asked why. Wheeler replied, "Because, they are all the same electron!"

Feynman immediately understood Wheeler's idea. Look at the Feynman diagram in Figure 20.5 and you'll see that the electron bounces backward into the past. Clearly the positron and the electron will have the same mass because they are the same particle. But suppose, in some distant past, the backward-moving electron scattered forward in time again. Then we would have two electrons coexisting that were actually the same particle. Maybe all electrons are connected in this way; there is only one electron, bouncing backward and forward in time.

Feynman says he dismissed the idea not because it was too crazy (nothing seems too crazy in physics), but because it seemed to predict that there would be equal numbers of positrons and electrons in the universe. If that's true, where are all the positrons? (It is characteristic of a great theorist such as Feynman that his immediate response was to see whether he could falsify the theory.) Wheeler responded that they might be hiding somewhere—maybe inside protons, for example.

This example also illustrates that Feynman and Wheeler, at least, took the backward-time interpretation as more than a gimmick. Feynman rejected the "all the same electron" speculation not because backward time wasn't real, but because of an experimental observation of unequal number of electrons and positrons in the current universe.

These days, Wheeler's idea seems more plausible. As I mentioned earlier, Andrei Sakharov showed (in 1967) that a tiny known difference between matter and antimatter ("CP symmetry violation") enables us to postulate a nearly but not quite equal number of particles and antiparticles in the early universe, and after most of them annihilated, to leave us with a current universe dominated by matter.

Maybe someday someone will organize a religion based on Wheeler's idea. Your soul, when you die, moves backward in time, scatters, and becomes a forward-moving soul, a different person. This happens many times. Maybe there is, indeed, only one soul in the universe. A nice aspect of this religion is that it doesn't require us to postulate the Golden Rule. In fact, the Golden Rule is an inevitable consequence. Whatever you do unto others, you are in fact doing unto yourself.

Can Humans Travel Back in Time?

Feynman's electron moving back in time doesn't seem directly relevant to a question that excites readers of science fiction: what about human time travel? In modern times (post H. G. Wells's 1895 novel *The Time Machine*), when science fiction tries to align itself with recent discoveries, such time travel is usually accomplished by one of two means: faster-than-light travel, or wormholes.

When, in *Superman: The Movie*, the hero discovers that Lois Lane is dead, he flies faster than the speed of light, moves backward in time, and acts to prevent her death—which, in his new proper frame, hasn't yet happened. But although his action is supposedly inspired by relativity, his achievement actually violates Einstein's equations. Recall that I showed that traveling faster than light can reverse the order of separated events. The tachyon gun can hit its target before being fired. But although this weapon makes causality ambiguous, there will never be any disagreement between observers. If Lois Lane dies in one reference frame, she dies in all reference frames, although at different times. So to save her, as in the movie, you have to postulate that something is wrong with relativity. But then, why travel faster than light? If you are not trying to be consistent with modern physics, why not just accept science fiction as valid and have Superman with his super brain build an H. G. Wells time machine?

For time travel, we could try to take advantage of the fact that a wormhole can connect a location in space-time with another location that has not only a different position but also a different, and perhaps earlier, time. Imagine a roll of (old-fashioned) movie film representing a space-time diagram. Now fold it over, and bring something that already happened in the past up to the present. Jump across and you are in the past. Character

Ellie Arroway flew through such a wormhole in Carl Sagan's (wonderful) novel *Contact*. If you would like to see a vivid depiction of a wormhole, watch Jodie Foster (playing Ellie) drop through in the (not as good) movie based on the novel. More recently, one of the key physicists who related wormholes to time travel, Kip Thorne, was an executive producer for a major science fiction movie based around this possibility, *Interstellar*. A wormhole was diagrammed in Figure 7.2 (page 90).

Time travel is so speculative that it is not normally considered a topic for professional publication, but there was a famous exception in 1988 when Thorne and two colleagues at Caltech published an article in the highly prestigious *Physical Review Letters* with the intriguing title, "Wormholes, Time Machines, and the Weak Energy Condition." (The term *weak energy condition* has to do with achieving a long-lived wormhole.) I'll refer to this as the "time machine paper." The abstract says,

> It is argued that, if the laws of physics permit an advanced civilization to create and maintain a wormhole in space for interstellar travel, then that wormhole can be converted into a time machine with which causality might be violatable.

This was a highly technical and carefully written article, and it is probably the work most responsible for the widespread assumption that time travel through wormholes is possible—even though the authors don't say that it is. They do suggest that a future highly developed civilization could, in principle, construct a wormhole connecting two different regions in both space and time. No practical method is proposed; the authors just argue that, with sufficient ability to garner huge resources of energy, nothing (well, almost nothing) in the known laws of physics prohibits doing it. Travel through their wormhole can go in either direction, so they argue that you can jump in and come out not only at a different place, but at a different time, even in the past.

It is the closest that serious physicists have ever come to suggesting a mechanism for a time machine. The authors conclude,

Consequently, at late times by traversing the wormhole from right mouth to left, one can travel backward in time . . . and thereby, perhaps, violate causality.

Violation of causality, as the tachyon murder paradox shows, implies negation of free will. As a vivid example, the authors of the time machine paper then bring up the subject of Schrödinger's cat! They say,

> This wormhole spacetime may serve as a useful test bed for ideas about causality, "free will," and the quantum theory of measurement. . . .
>
> As an infamous example, can an advanced being measure Schrödinger's cat to be alive at an event P (thereby "collapsing its wave function" onto a "live" state), then go backward in time via the wormhole and kill the cat (collapse its wave function onto a "dead" state) before it reaches [the alive cat event]?

Nowhere in the time machine paper is there any discussion of the arrow of time, of the fact that the arrow has to continue to point along the direction of the wormhole path even as it reaches the past. Time travelers passing through the wormhole must do so without the arrow reversing, so that they can reach their destination while experiencing normal time progression. That's a critical but unaddressed issue.

I would argue that true time travel, if possible, would mean that the traveler's concept of *now* must go from the present to the past. The time machine paper does not discuss the issue of what motion along that path would do to the traveler's notion of *now*. The authors say that the wormhole allows them to draw a "closed timelike curve," physics jargon for a path that contains a section that takes you back to your past. But could a person travel along that path and still experience progressing time, while retaining memory of what has become the future? I can always reverse an arrow on an electron and call it a backward-time-traveling positron, but is that the same thing as the backward time travel of H. G. Wells?

Another catch in the paper is that the wormhole is described as so unstable, so short-lived, that a person would not have sufficient time to travel through before the wormhole disappeared. There is a loophole: if physicists and engineers can figure out how to impart a "negative-energy density" to a large region of space, then the wormhole might endure. No way to do that is known, but nothing in physics absolutely rules it out, we think. With this requirement, however, the entire demonstration of the feasibility of stable wormholes has collapsed, independently of the other objections. It has become speculative, requiring new physics. The authors are clear about this. They state, "Whether wormholes can be created and maintained entails deep, ill-understood issues." The existence of such wormholes is reminiscent of the possibility of tachyons: just because nothing in our current physics theory rules them out doesn't mean that they do exist.

Finally, even if the progression-of-time issue could be answered, and if the required negative-energy fields could be discovered, there is still the issue of causality and free will. The time machine paper does discuss this, if only to point out that there is a *reductio ad absurdum*—using the cited example of Schrödinger's cat. Closely related is the *grandfather paradox*, in which you go back in time and kill your grandpa. Since no grandpa would mean no you, how could you do that if you didn't exist? One possible answer is that you don't have free will, so even if you went back in time you could not kill your grandfather. And the fact that you were eventually born shows you didn't do that.

One way to keep free will is to postulate some kind of cosmic "censorship"; that is, you can go back in time but can't change what has already happened. That is what happens to Claire in the novel (and TV series) *Outlander*; she uses knowledge of the future to change things, but circumstances always conspire such that what she does has no effect: (spoiler alert) she thinks she kills her husband's ancestor, only to find out that her husband is the descendant not of that ancestor's blood child, but of an adopted child. In *Back to the Future*, time travel into the past can and does alter the present, with humorous consequences that, for an unexplained reason, do not invalidate the memories of the traveler.

More important, in my mind, is the question, What is the value of going back in time if you can't change things?

As Sarah Connor says in another time travel movie, *The Terminator*, "God, you can go crazy thinking about all this."

Physics analyses of time travel assume the standard fixed space-time diagram. Indeed, that is the current way that most physics is computed, and the way the physical world is represented, but we all know that it is not the world of our experience. If everything in the future and the past is already determined, what would be the value of time travel? The standard space-time diagram has no way to indicate *now*, and in time travel, it is the *now* that we want to change.

Wormholes are an interesting phenomenon for physics calculation, and they readily gain attention from the science fiction (and cartoon) community. Wormholes may be a way of changing positions with equivalent speeds that exceed lightspeed. But if we *really* want time travel, we have to understand the meaning of *now*.

PART IV

PHYSICS AND REALITY

21

Beyond Physics

*An exploration of knowledge that is meaningful
but not experimentally measurable . . .*

Who steals my purse steals trash—
But he who takes my time takes my life.
— *Apologies to W. Shakespeare*

Einstein was in awe, not only of physics, but also of his own contributions. Why had he succeeded? In 1921 he wrote,

> How can it be that mathematics, being after all a product of human thought which is independent of experience, is so admirably appropriate to the objects of reality?

Actually, it isn't. Nobody has good equations to describe living organisms, the process of thinking, or even economic interactions among people. Oh, that's not physics, you might say. True, but beware of the potential for tautology.

Physics is arguably that tiny subset of reality that is susceptible to mathematics. No wonder physics yields to math; if an aspect of existence doesn't so yield, we give it a different name: history, political science, ethics, philosophy, poetry. What fraction of all knowledge is physics? From an information theory perspective, the answer is, "very little." What fraction of what you know that is *important* is physics? I imagine that even for Einstein that number was tiny.

Limitations of Science

When I was a sophomore at the Bronx High School of Science, a senior (who was dating my sister) gave me a paperback book, *The Limitations of Science*, by John William Navin Sullivan. I still have the copy I marked up—Mentor Edition, 50¢, ninth printing 1959—of the original 1933 classic.

I hated that book. It upset my belief that science was the ultimate means to knowledge, that it was the arbiter of truth, that it could allow us to see clearly into the future. I was so disillusioned that I thought maybe I should major in English, not physics. Yet I read every word and marked a few dozen paragraphs as particularly bothersome or important. One section I underlined, on page 70, said,

> The principle of indeterminacy is founded on the fact that we cannot observe the course of nature without disturbing it. This is a direct consequence of the quantum theory.

This was my first encounter with the Heisenberg uncertainty principle; when Sullivan wrote this book, that principle had not yet acquired its modern name. The phrase "without disturbing it" would now be more precisely written as "without collapsing the wave function." Science could not make predictions. It could only estimate probabilities. I was disenchanted.

Little did I realize at that time that what bothered me was the same thing that had bothered Einstein. He could not abide the concept that physics was *incomplete*, that it was not a full description of reality, that the past did not completely determine the future.

At the time Einstein was struggling with these issues, a recent development had set the tone—something perhaps even more surprising than the limitations of physics. Einstein knew that all mathematical theories were *incomplete*. That fact had been discovered and proved in 1931 by his friend at Princeton, Kurt Gödel.

A Shock from Gödel

Gödel had proved a mathematical theorem that traumatized not only mathematicians and physicists, but also philosophers and logicians. It is not mentioned in Sullivan's 1933 book, possibly because it was still so new that few people understood it, maybe because few believed it and many who did hoped it would be shown to be flawed. Or maybe Sullivan didn't consider math to be science; in much of Europe, math is considered a liberal art, along with music and philosophy. Time has passed, and now Gödel's theorem is considered both fascinating and enormously important, widely regarded as the greatest mathematical achievement of the twentieth century.

Gödel's theorem can be stated in a deceptively simple way: *All mathematical theories are incomplete.* What this means is that any mathematical system you devise will have truths that cannot be proved—indeed, cannot even be identified as truths.

Gödel didn't prove that math is incomplete, but only that any one set of definitions, axioms, and theorems is necessarily incomplete. For example, there are some theorems you might not be able to prove using real numbers, such as the possibility that π^e is irrational. (Here, π is the ratio of a circle's circumference to its diameter, and e is the base of natural logarithms.) Yet if you extend the number system to include imaginary numbers, it might be possible to prove this theorem. (In fact, we don't know whether π^e is irrational or not; I cite it only as a possibility to illustrate Gödel's result.) But then, once you have extended your math, yet another theorem is bound to exist that is true but unprovable.

Another possible example is the conjecture by German mathematician Christian Goldbach that every even number can be written as the sum of two prime integers. This idea, too, has not been proved, and there is no empirical way to determine its truth. It may be an unprovable theorem, given today's mathematics. (If you think you can prove it, please send your proof to a professor of mathematics, not to me.) But this conjecture might yield to proof someday, or maybe in a future extension of math.

The reason you can't identify theorems as true but unprovable is simple: if you could identify them, that would be a proof of their veracity. Many theorems can be disproved with one counterexample. That's not possible with the Gödel theorems.

Because modern physics uses mathematics as its principal tool, any physics theory will necessarily be incomplete. There will be true statements that cannot be proved, or shown to be true. Stephen Hawking laments this fact but takes comfort in the recognition that you can address any unknown by developing a more complete theory, by adding more postulates or "principles." He reasons, from Gödel's theorem, that all theories we have so far (and, he would certainly agree, will have in the future) are incomplete. He humorously concludes that there will always be jobs for theorists.

Gödel's theorem inspires us to wonder about the completeness of physics—not of any particular theory, but of physics itself. Are certain aspects of reality, in addition to those affected by the uncertainty principle, beyond the reach of physics? Once you start thinking along these lines, you discover that many aspects of reality not only are untouched by current physics, but also appear to be untouchable by any future physics advances. One example is evident in the question of what something *looks like*.

What Does Blue *Look Like*?

When you see blue and I see blue, are we seeing the same color? Or could it be that when you see blue, you are really seeing what I see when I see red?

This question bugged me in fifth grade. My teacher wasn't helpful. "Of course we all see the same," she said. I didn't give up. In ninth grade, my science teacher seemed to know a lot, so after class I asked him. He told me that the signal goes to the same part of the brain in all people, so of course we all see it the same. I didn't think he had answered the question. I also learned to beware of the phrase "of course."

How could I formulate the question in a more compelling way? That proved to be tricky. Some people seemed to know what I meant; others dismissed it as a nonsense question. I now know that many of the world's great philosophers were bothered by the same issue. The problem could be summarized by the distinction between the brain (the physical object that does the thinking) and the mind (the more abstract concept of a spirit that uses the brain as its tool). The brain-mind distinction was one of a class of problems labeled *dualism* that dates back at least to the ancient Greeks.

Here is a simple experiment you can do to clarify the color question. Keeping both eyes open, look at a colored object; then alternatively cover your left eye and your right eye with your hand. Are the colors exactly the same? For older people, usually not; the lens of the eye is getting a bit discolored, each eye a little differently, and those changes alter the perception. It's like looking through glasses with a different tint over each eye. My ophthalmologist tells me that many people see colors slightly differently in their two eyes. If red looks slightly different to your two eyes, could it look completely different to another person? (Switching physical eyes does not address the question.)

I have a condition called *diplacusis binauralis*. It is the minor annoyance that, for the same frequency (say, from a tuning fork), I hear different tones in each of my two ears. The main annoyance is to my children, who for a long time complained that I could not carry a tune. I finally figured out how to sing by matching the tune differently in each ear simultaneously.

These are small effects, but there is no reason why they couldn't be large. Maybe my blue really is your red.

In 1982, Australian philosopher Frank Jackson posed my childhood color question in a way that I find particularly compelling. He created the story of Mary, a brilliant scientist who had been raised indoors in a colorless environment, with nothing to look at that wasn't black, gray, or white. She only read books without color pictures and watched black-and-white television.

The Exploratorium museum in San Francisco has a wonderful room that simulates a colorless environment. The room is lit with a nearly monochromatic light—single frequency, one color only, yellowish, from low-pressure sodium lamps. (You can buy one and try it yourself at home; don't get the high-pressure lamp, which emits a range of colors.) The Exploratorium's room is full of objects that in white light would be colorful: fabrics and montages, even a jelly bean dispenser, but none of the color is visible, only shades of yellow: bright yellowish, gray yellowish, and dark yellowish. And as you stay in this room, even the perception of yellow fades, just as you sometimes forget that you are wearing tinted sunglasses after a few minutes. Your eyes "get used to it" and you see only black and gray and white. But flashlights are available, and if you shine one on the jelly beans, you are dazzled by the burst of colors. (If you take a child to the Exploratorium, be sure to bring a quarter for the jelly bean dispenser.)

Jackson's imaginary brilliant scientist Mary, in her black and white and gray home, grows up normally, except for the absence of color. She reads about color in her physics books. She wonders what living in a world with color would be like. She finds the theory of the rainbow to be elegant, beautiful (in the physics sense), but she ponders, what would a rainbow actually *look like*? Would the beauty be different from the physics beauty?

Ultimately, Mary becomes a "brilliant scientist," a master not only of physics but also of neurophysiology, philosophy, and any other discipline you might like to throw in. (Remember, this is an imaginary story.) She understands how the eye works—how different frequencies of light stimulate different sensors in the eye and how the eye does some initial processing and then sends signals to different parts of the brain. She knows all about this, but she has never experienced it herself.

Then, one day, Mary opens the door and walks outside into a full-color world. What will her reaction be when she finally sees a rainbow? (Remember, this is a thought experiment; we aren't worrying about whether all those years without color atrophied her visual ability.) When she looks at the sky and the grass and a sunset, will she say, "Oh, this is exactly what I expected, from the science I studied"? Or will she say, "Wow! I had no idea!" Jackson asks, "Will she *learn* anything or not?" And if she does learn something, what will it be?

My answer to Jackson's question is yes, she will learn something: she will learn what red, green, and blue *look like*. But if someone else— you?—answers Jackson's question by saying that she will learn nothing, I'll be hard-pressed to convince you that you're wrong. Either you know what I mean or you don't. I can't use physics or math or any other quantitative science to explain it to you. And likewise, you will have a hard time convincing me that I am wrong. You may conclude that I am not open-minded, that I am not objective, that I don't listen to reason, that I am not being scientific. I claim that I *know* what I am saying is true. It is not my *opinion*, not my *belief*. I know what I mean, and it is true! There is additional knowledge about color that Mary learns only when she sees it herself. She learns what it *looks like*. You say, nonsense; she has learned nothing. There is no way for you and me to reconcile our differences.

What is it that does the seeing? If free will exists, what exercises it? What experiences *now* and differentiates it from *then*? Is something hidden deeper in the brain, or is it beyond the brain? To sharpen the question, consider the teleportation of James Kirk, captain of the starship *Enterprise*.

Beam Me Up, Scotty

One of the classic sound bites from the *Star Trek* series is the iconic phrase, "Beam me up, Scotty." When Captain Kirk used these words,*

---

* Trekkies know that this exact phrase was never used in the original series, although Kirk did once say, "Scotty, beam me up."

his engineer, Scotty, would activate the teleporter that would cause his body to vanish (disintegrate?—we never know for sure) and then reappear (be reassembled?) at a different location. It was the ultimate in high-speed and convenient transportation, and in *Star Trek* it sped up the story line.

How did it work? Well, of course, it didn't; this is science fiction. But when I watch science fiction, I always try to reconcile it with physics. In this case, it wasn't too difficult. Think of Kirk as a quantum wave function. The teleporter simply had to "clone" this wave function, to create an exact duplicate. Was the duplicate made with the original molecules? It hardly matters; carbon atoms are all identical in modern physics; likewise for all electrons, for all oxygen atoms, and so forth. Remember the Feynman diagrams (Figures 20.3 and 20.4), which operate simultaneously when an electron is deflected by a positron. The fact that both diagrams contribute to every scatter means that the emerging electron is both the original electron and a newly created one, simultaneously. Another example of the challenges of thinking about identical particles is this: you currently have few of the same molecules you had in your body when you were a child; most of them were replaced, but you still feel like the same person.

Several theorems in modern quantum physics, it turns out, show that such cloning is possible, in principle, *only* if you destroy the original. One is called the *no-teleportation theorem*, but despite the negative in its name, it doesn't rule out the *Star Trek* version; it simply states that you cannot teleport by converting the wave function first into a classic set of measurements and then back again. Another theorem is the *no-cloning theorem*, but it doesn't mean you can't clone; it means only that you can't make an exact copy without destroying the original. So although we may not know how to do a *Star Trek* teleport, nothing currently known in physics rules it out.

Suppose we figure out how to teleport along the lines of *Star Trek*. Would you allow yourself to be so beamed?

I wouldn't.

Why? I worry that the new person appearing at the end of the beam might not be me. I accept the premise that the created human would have all of my memories, all of my characteristics, all of my foibles and loves and likes, and would be indistinguishable from me by any physical measurement. But would he be *me*? Do you see why I'm concerned? Is an exact clone exactly the same as me? Certainly, physics couldn't tell us apart. But is there a reality beyond physics? Put into the old language of religion, how do we know that my *soul* would be transported along with my body?

Science fiction boasts a conceit that the human body can be duplicated, along with memory, but that the person so cloned is indeed different. In the books and movies that depict this scenario, adults have a hard time telling the original from the copy, but children and pets can see the difference easily. And like Cassandra in the Greek legend, nobody believes them, even though they speak the truth. Such has been the case in soul replacement movies, from *Invasion of the Body Snatchers* (1956) to *Invaders from Mars* (1986). The clones typically try to convince the noncloned that cloning is a wonderful thing to do. But as we watch the movie, we know that it isn't.

Am I transported when my quantum wave function is cloned? That's truly a nonsense question. Right?

What Is Science?

What distinguishes scientific knowledge from other kinds? I think that the defining essence is that science is that subset of knowledge on which we can aspire to universal agreement. Science has the means to resolve disputes, to determine what is right and what is wrong. You and I may never agree on whether chocolate is delicious or yucky, but we know we can ultimately reach agreement on the mass of the electron. We may never agree on the best form of government, on economic systems, maybe not even on justice and ethics, but we can expect to agree on whether relativity is correct, and whether $E = mc^2$.

When I see blue, do you see blue? That is not a scientific question. Does that make it invalid? This issue is related to the difference between the brain and the mind. Is there something beyond the brain, something behind the circuitry, something more than a physical, mechanical set of atoms, something that can not only see, but knows what a color *looks like*? I can't prove to you that such knowledge exists. I can only attempt to persuade you.

The problem is similar for the irrationality of $\sqrt{2}$, for the fact that it cannot be written as the ratio of integers. The proof I give in Appendix 3 is based on reaching a contradiction, an approach that can't be deduced from prior math; it has to be assumed; it has to be accepted as a postulate. The situation is similar with mathematical induction; there is no proof that the method is valid. It must be taken as a separate assumption. And there is the important—but a bit more obscure—*axiom of choice*, a key concept in mathematics. Even our ability to *choose* is not self-evident. And, in fact, it may be wrong if we are really just machines driven by external forces, abetted perhaps by a God who throws dice.

In discussing what colors "look like," we are departing from the rules that Newton implicitly followed in his studies of physics. Some would complain that we are drifting from science to semantics or—even worse—philosophy. We are not discussing issues that have real meaning or interesting content. Please, you say, just define with precision what you mean by what a color *looks like*, and then we can determine whether it is universal.

Plato argued in his dialogue *Meno* that there is knowledge that cannot be obtained by physical measurement. Plato was referring to *virtue*, among other things—a concept that many scientists today would dismiss as "unscientific." Virtue, they might argue, is a set of conducts optimized by whatever behavior leads to survival of the fittest. Plato demonstrated his contention of intrinsic knowledge by never (well, rarely) having his protagonist Socrates give his own opinion, but instead by having Socrates ask questions to draw out knowledge that he claimed already existed within the mind of his subject. Imagine that instead of giving a proof that $\sqrt{2}$ is irrational, I simply ask you questions and, in doing so, lead you to the proof yourself. That's the essence of the Socratic method. Then I could claim, as does Socrates, that the knowledge was already in your brain and only had to be teased out.

All of mathematics is knowledge that is outside of physical reality. That is what bothers many people about the subject, and it is the cause of much math phobia. Empirically, we can only show that some rules of math are approximately true. Is the Pythagorean theorem exact? Or is the largest angle in a 3-4-5 triangle not 90 degrees but really only 89.999999 degrees? How do you know? Not from physics; not from measurement. (And in curved space, it turns out not to be 90 degrees.) Mathematics investigates truths not by experimental tests, but only by self-consistency. You can postulate that two distinct lines passing through a point will never again meet, or you can postulate that they will. The first is part of the foundation of Euclidean geometry; the second is true in a closed, curved space-time of general relativity.

According to legend, the Pythagoreans were so upset by the discovery that $\sqrt{2}$ is irrational that they threw Hippasus, the man who discovered it, overboard from a boat. (The modern metaphor is "throw him under a bus.") Hippasus's proof may have been similar to the one I give in Appendix 3, but there are nice alternative proofs, one based on geometry.

According to another version of the legend, the Pythagoreans considered the discovery of the nature of $\sqrt{2}$ to be so profound that it became the foundation of their religion. In that story, they threw Hippasus overboard to punish him for having revealed this great secret to outsiders. But it is certainly true that the Pythagoreans had uncovered in this theorem the deep truth that there is knowledge that exists outside of physical reality, a truth so astonishing that they revealed it only to those sworn into the Pythagorean faith. Hippasus had discovered that nonphysical truth, truth that defies physical verification, does indeed exist.

22

Cogito Ergo Sum

Does now exist in the brain? Or only in the mind?

Come, let me clutch thee.
I have thee not, and yet I see thee still.
Art thou not, fatal vision, sensible
To feeling as to sight? or art thou but
A dagger of the mind, a false creation,
Proceeding from the heat-oppressed brain?
— *Macbeth*

We take this truth to be self-evident: *If it isn't measurable, then it isn't real.* That "truth" is not provable, of course, any more than are the rights proclaimed in the Declaration of Independence. But it is not a hypothesis, and certainly not a theory; it is more like a doctrine, a thesis figuratively nailed to a physics department door, a dogma that, given faith, will lead you to mastery of the physics world. Philosophers call this dogma *physicalism.*

Please don't misunderstand what I am saying. Physics itself is not a religion. It is a rigorous discipline, with strict rules about what is considered proven and unproven. But when this discipline is presumed to represent all of reality, it takes on aspects of religion. Not only is there no logical imperative between physics and physicalism, but there is no logic whatsoever linking them. The dogma that physics encompasses *all* reality has no more justification than the dogma that the Bible encompasses all truth.

Physicalism

Physicalism is illustrated in the quote I cited in Chapter 1 from philosopher Rudolf Carnap criticizing Albert Einstein's drift toward nonphysics beliefs: "Since science in principle can say all that can be said, there is no unanswerable question left." That is self-evident, right? When you read it, did you accept it as a well-established truth?

What do colors look like? That's not a physics question, so physicalists wouldn't tolerate it. When you see blue, is it the same as when I see blue? That question is nonsense, meaningless. You cannot lay out a procedure that will enable the answer to be tested, and therefore it is incapable of having its truth evaluated. To physicalists (perhaps I should capitalize the term to emphasize its similarity to a religion), just asking such a question throws your judgment into doubt. Just asking what a color *looks like* makes physicalists wonder whether you're drifting away from physics, losing your discipline, slipping toward scientific apostasy.

Physicalism reaches its extreme when it asserts that nonquantifiable observations are illusions. You and I think we know that time flows, but it really doesn't. Since it doesn't exist in current physics theory, since it doesn't appear on a space-time diagram, then it isn't real, since the current physics structure, even if it doesn't answer all questions, does cover all of reality.

Physicists usually include mathematics as a science because of its rigorous discipline. Not everything need be tested empirically; we can also test its *consequences*. We know that the square root of 2 is irrational; that is, that it cannot be written as the ratio of two integers. That claim could be falsified, though only within the abstract but self-consistent realm of math, if we find integers whose ratio yields $\sqrt{2}$.

Physicists do use quantum amplitudes and wave functions, things that are not measurable, but they are embarrassed about them and apologetic. They hope that someday they'll be able to eliminate them. In the meantime, they avoid talking about their interpretation. Physics is validated, despite its failings, by the miracles it produces: radios, lasers, MRIs, television, computers, atomic bombs, and so on.

Atheism, by itself, is not a religion. It is a denial of a particular kind of religious belief, *theism*, that states there is a God who will reward worship by helping your football team win, or your army, or by curing your cancer. Atheism is only a rejection; it doesn't become a religion itself until it incorporates a positive faith, such as physicalism, the belief that all of reality is defined by physics and math, that everything else is an illusion.

It is remarkable how often you run into the phrase "Science says..." to support an idea that actually has no foundation in science. It is often physicalism in disguise. "Science says we have no free will." Nonsense. That statement is inspired by physics, but it has no justification in physics. We can't predict when an atom will disintegrate, and the laws of physics, as they currently exist, say that this failure is fundamental. If we can't predict such a simple physical phenomenon, then how can we imagine that someday we will be able to show that human behavior is completely deterministic? Yes, on average we know radiocarbon atoms will decay in several thousand years, and on average we expect that humans will make decisions that enable them to procreate more humans, but even if you accept that minimal scientific conclusion, it leaves lots of room for decisions based on ethical and empathetic values. Science doesn't "say" that we will be able to understand human decisions without including human free will.

Richard Dawkins, the author of *The God Delusion*, is the "primary atheist of the world," according to astrophysicist–author–*Cosmos* star Neil deGrasse Tyson. I love Dawkins's books on science, and he properly and effectively attacks many of the counterfactual claims of religious sects. His criticisms of organized religions are often valid, but because nonphysics knowledge is responsible for much evil, he seems to think that it is all balderdash. Dawkins makes a fundamental error in his unstated but implicit postulate that logic requires us to ignore nonphysics reality. A corollary of this is the mistaken belief that mastery of physics is incompatible with worshipping God. I give some counterexamples in Appendix 6, deeply religious statements made by some of the greatest physicists of all time.

In his 2006 book, *The God Delusion*, Dawkins says, "I am thrilled to be alive at a time when humanity is pushing against the limits of understanding. Even better, we may eventually discover that there are no limits." Dawkins hopes that there are, indeed, no limits to the capability of science, but it seems to me it is more than a hope and has become his belief. It is the foundation of his religion. It is based on the success of science in explaining *so much*, and his belief that therefore it will explain everything. His optimism reminds me of that of the ancient Greeks, who expected that all numbers could be written as ratios of integers. He will be disappointed. The limits of physical knowledge are severe and obvious. Several examples that I've already given make it clear to me that physics is incomplete, incapable of describing all of reality.

Moreover, Dawkins's faith in the supremacy of logic neglects the stark limitations discovered by Kurt Gödel. As noted earlier, Gödel showed that all mathematical systems have unprovable truths, truths that cannot be addressed or tested through the use of logic. So Dawkins's approach to reality, accepting only truths that are logically demonstrable, is patently wrong even within the simple and clean realm of mathematics.

Empathy

Have you ever tried to imagine what it would be like to be someone else—a friend, a spouse, or a famous person (Joan of Arc, Albert Einstein, or Paul McCartney)? When you imagine that, do you assume that you forget all your own memories and just are that other person, seeing the world through her or his eyes? This capability of the mind is thought to be the source of empathy. Inability to do this is the fundamental dysfunction of sociopaths. What is it that you are imagining when you imagine being someone else? What part of you is transferred? Not your own feelings or memory or knowledge. You are trying to see the world as it is seen by that other person. What does that mean?

For lack of a better term, let's refer to the thing that you imagine transferring to the other being as your *soul*—the same thing that just might not go along when you get beamed up by Scotty. I hesitate to use the word *soul* because it carries a lot of baggage from its use in religion: immortality and memory independent of the body (will you recognize your parents' souls when you die?), the thing that is punished when you sin. So I was tempted to call it your "quintessence" (a term already appropriated in cosmology theory), or your "anima" (too closely associated with hypnotism), or your "spirit" (associated with enthusiasm for sports), or the French word *esprit*; but let's just stick, for simplicity, with *soul*. Does it exist? Is it real?

The soul appears to be impossible to detect with physics, although people have searched for it in the physiology of the brain. It is often confused with "consciousness," probably because consciousness is more amenable to physicalism. The soul-consciousness difference is analogous to the mind-brain difference.

I recall that in fifth grade my teacher said she was going to teach us how we *see*. (This was the same teacher I later asked about colors.) I was excited. This was something I had deeply wondered about and wanted to understand. That afternoon, she began. She pulled down over the blackboard a diagram of the eye. I had seen that in the science books I had borrowed from the Mott Haven Branch of the Bronx Public Library. (You can't ask books questions.) Nothing new yet. She traced the rays. Yes, I knew that too. Light went through the lenses, was focused on the retina, and was turned into electricity. I had read about that. Pulses went to the brain. The brain knew where each signal came from so it could reconstruct the image. The retina image was upside down, but the brain inverted it. OK, here we go! This was the moment when my questions would be answered! My concentration doubled. (Yes, this is a true story; I really was on the edge of my seat.) But instead of giving the explanation, she said, "Now let's talk about the ear and how we hear."

Eruditio interruptus!

I sat back terribly disappointed. I had read the science books, but they always stopped at the brain. I wanted to know how *I* saw, how the signal went beyond my brain, to that place where I could see the what the color of blue *looked like*. As I mentioned earlier, I went up to my teacher and asked her afterward, and she didn't seem to understand what my question was about. The signal goes to the brain; that's it.

What does all this have to do with the mystery of *now*? As long as we think we are nothing but machines run by a fancy multitasking computer, the issue of *now* is irrelevant. It has no meaning, unless perceived by that same thing (the soul?) that looks at the signal in the brain and sees what blue looks like. That doesn't mean that *now* doesn't have a physics origin; I think it does.

The body processes signals, but the thing that *looks* is what I refer to (for lack of a better term) as the soul. I know I have a soul. You can't talk me out of that. It's that thing that goes beyond physics, that is beyond the body and past the brain and sees what things and colors *look like*. I don't understand the soul. I doubt that my soul is immortal—but having had children and grandchildren, I feel more and more every day that there is a kind of immortality that I achieve through them. Do they have souls too? Yes, that is obvious to me, yet I can't explain how I know. I feel that the clear perception of another person's soul is the essence of empathy, of love. How could you possibly do harm to another person when you are aware of that person's soul?

Yet I know people, sociopaths, who act as if they don't share such perceptions. They treat others as if they were machines. Doing harm to another person is no more to them than would be discarding a bicycle. They lack empathy, that ability to put themselves in the shoes of someone else, to recognize that another person has a soul. I take some comfort in the fact that such people have been recognized and categorized by psychologists, that they are outliers, not the majority of humanity.

People without empathy are often depicted in books and media. When young Jimmy in the 1956 movie *Invasion of the Body Snatchers* says, "She's not my mother!," what he may be sensing in the person who looks and acts like his mom is her absence of empathy. In the 1998 movie *Dark City*, the aliens have built an entire planet to run experiments solely to figure out what makes humans human. At a climactic moment, the protagonist, John Murdoch, points first to his brain and then to his heart, and proclaims that they were looking in the wrong place; the human essence is found not in the logical thinking of the brain, but in the empathy represented by the heart.

In elections in the United States, I sometimes feel that what the voter is most interested in is whether the candidate has empathy—for the voter, for the poor, for all others. Policy issues come second. The US voter does not want to elect a sociopath. Yet sociopaths can be very successful leaders, as evidenced by Mao, Stalin, and Saddam Hussein.

If you tell me that I don't have a soul, that it is an illusion, that you can teach a computer program to act as if it, too, has a soul, I conclude that you don't know what I'm talking about, just like my fifth-grade teacher. My soul is blatantly obvious to me, even though I have a hard time expressing what I mean by it. Using Augustine's words (he was referring to the flow of time), "If no one asks me, I know; if I wish to explain, I do not know." Anytime I think about it, I feel a sense of wonder. It is my primary religious revelation. I suspect it was such an experience that Einstein referred to when he said, "A person starts to live when he can live outside himself."

Many unanswered questions remain. Do animals have souls? I don't know. Virtually all the dog owners I have known over the years believe that dogs do. I once got to spend two hours (spread over two days) within a few yards of two families of wild mountain gorillas in Rwanda and I came away convinced that they had souls. They appeared to be wild, big, strong, hairy humans.

Cogito Ergo Sum

In 1637, Descartes wrote, "I think, therefore I am"—his concise refutation of the philosophy that life is an illusion, his refutation of the claim that we don't even exist. Descartes's statement has been discussed, debated, and dismissed. Certainly, if you insist on a strict definition of all the terms, it is either trivially true or trivially false. But why would Descartes spout trivia? And why has his phrase stuck with us so strongly?

I think his famous phrase can best be interpreted as a refutation of physicalism. The original words were written in French, not English or Latin, and they were, simply, "*Je pense, donc je suis*." Classical philosophers interpret *pense* as the physical act of thinking, signals moving about in the brain. This classical interpretation could be equally applied to a modern computer. But I find Descartes's statement most compelling when I interpret *think* as referring to the action not of the brain but of the mind, the spirit, the thing that sees what color looks like, that hears what music sounds like, that exercises empathy. Science could describe existence as abstract, with reality an illusion, but Descartes *knew* that it wasn't so. Although Descartes wrote in the 1600s, the issue is still alive today; in physics, it is indirectly related to the *holographic principle*, a reinterpretation of reality that these days is a favorite of many string theorists.

Physicalists have a practical reason for wanting to deny nonphysics knowledge. Once you allow it, you open the floodgates of spiritualism, pseudoscience, and religion. You lose all control; anybody can say anything as long as it doesn't contradict observation. Math may be nonphysical, existing only in the mind, but at least math has strict discipline, rules and procedures, ways to falsify incorrect propositions. Any talk of the soul, however, leads to "truths" that have no requirements for self-consistency, cannot be tested, and are therefore suspect, possibly a waste of time, distracting and misleading, perhaps even dangerous.

In 1996 there was an ethical debate over the issue of a sheep that had been cloned. Her complete genome had been used to engender a second sheep, Dolly, as identical in physical makeup to the donor as are two identical twins to each other. Next, people feared that wealthy people would be making clones of themselves.

So what? What is there to fear? Well, many people were disturbed by what they perceived as the ethical implications. As with many other new issues in science, from vaccinations to birth control, the perpetrators were accused of "playing God." One issue was whether the cloned individuals would have souls. If they didn't, then could they be enslaved, just as we today enslave horses, dogs, automobiles, and computers?

People said we needed to discuss the full ethical implications of cloning before we allowed science to proceed. How long do you think that would take? For some reason, the comparison to identical twins (who everyone accepts as having separate souls) was rarely brought up. (The concept of the evil twin dates back to Zarathustra.) I mention cloning because it, too, shows that the sense of a soul is widespread. Many atheists accept the concept of the soul; they just deny the existence of a favor-bestowing God. It is primarily physicalists who reject the reality of the soul.

23

Free Will

*One important piece is not yet in place. It's on the edge of
the quantum physics section of the puzzle, and it will prove key
in giving now its special meaning . . .*

Just as the constant increase of entropy is the basic law of
the universe, so it is the basic law of life to be ever more
highly structured and to struggle against entropy.
— *Vaclav Havel*

Do you have free will?
I think I do, but I'm not completely sure. At least some of my
free will could be an illusion. My first doubts came in 1980. Two years
earlier, my wife, Rosemary, and I had had our first child. What to name
her? It felt like one of the biggest decisions in our lives. We wanted a
name not too common but not too unusual, one with personal significance but not too personal; it was to be her name, not ours. We read
the books with hundreds of names and meanings behind the names,
threw away the books, considered the nicknames we wouldn't like and
couldn't control, and suddenly discovered that *Elizabeth Ann* was our
choice.

I'm sure we were influenced in part by our admiration for powerful
and productive Queen Elizabeth I of England. Rosemary had taken a
course in Shakespeare from Professor Hugh Richmond, and I had
audited every lecture; Richmond and Shakespeare had both deeply
influenced our lives. Why the second name Ann? It just sounded right.

Only decades later did I notice the similarity of *Elizabeth Ann* to *Elizabethan*, the great era presided over by Elizabeth I. No wonder it sounded right. Had we named our daughter after a queen or after an era? We liked the nicknames. Elizabeth, Liz, Betsy, and Bess, all went together to see a bird's nest . . . . It was a very personal choice.

Or so we thought. Two years later, in 1980, I read a magazine article about popular names. For 1978, in northern California, the most popular name for a girl happened to be Elizabeth.

What is free will? Is it the ability to choose to do something without being influenced? Why would I ever want to do something without being influenced? But if my actions are determined by outside forces—the people who happen to be my parents and teachers and friends and colleagues, the books I happen to read, the experiences I happen to have—then does that make me into nothing more than a physical particle, passively pushed around by other particles, responding to forces just like the planets respond to the gravity of the sun, orbiting in a predetermined path, with only a delusion that I am acting on my own? Am I simply a wood chip caught in a complicated machine, bouncing around as the gears turn, confusing my rapid action with my importance?

When classical physics reached its zenith in the late 1800s, it appeared as if physics would soon be able to explain everything. True, there were a few problems—issues with the measurement of the absolute velocity of the Earth, and some unexplained aspects of thermal radiation. These minor unexplained issues turned out not to be minor; they led eventually to relativity and quantum physics, respectively.

Classical physics, even including relativity theory, was deterministic. The universe was causal. The past determined the future, completely. That suggested that even behavior is determined, in principle, by previous events. Later developments in chaos theory implied that we might never know the past well enough to predict the future, but that didn't change the determinism argument. All action, including that of humans, was predestined; the Calvinists were right. It was hard for philosophers to disagree with the discoveries of physicists. The rapid development of physics lent credence to the philosophy (religion?) of physicalism. In fact, the physicalist denial of free will may have reinforced the growing conviction that criminals were the victims of their upbringing and that punishing them for their actions was unfair. Society was responsible for all wrongdoing, not individuals. That's an odd conclusion that seems to attribute free will to society (to cope with wrongdoing) just as it denies it to criminals.

But the very premise on which this philosophical conclusion was based turned out to be false. All we need to demolish that argument—the claim that physics has shown free will is an illusion—is to demonstrate that physics is not causal, that future behavior of particles depends on more than just past experiences. I've demonstrated that myself, in my own lab.

Back in My Laboratory . . .

From the time of Newton to the time of Heisenberg, it was implicitly assumed that knowledge of initial conditions would determine the future of a physical system. Yet we now know that two objects, objects that are completely identical, identical in every way, can behave differently. Two *identical* radioactive atoms decay at different times. Their future is not determined by their past or by their condition, their quantum physics wave function. Identical conditions do not lead to identical futures. Causality affects the average physical behavior but not specific physical behavior.

Let me make this case in what I consider to be the most compelling way: by experiments and measurements I've done myself. Working in experimental elementary-particle physics in the late 1960s, every day my colleagues and I used the Lawrence Berkeley Laboratory's Bevatron*

---

* At the time, the laboratory was called the "Lawrence Radiation Laboratory at Berkeley." It was later renamed the "Lawrence Berkeley Laboratory." Now it is called the "Ernest Orlando Lawrence Berkeley National Laboratory," or "Berkeley Lab" for to smash protons into other protons. Many of these collisions produced two or more particles called *pions* (short for *pi mesons*). Here's a key fact: I could determine, experimentally, that all the pions from a single collision, if they had the same electric charge, were identical. I mean *really* identical, identical down to their deepest quantum core. These particles had identical quantum wave functions. They were identical in the same sense that the incoming electrons in the Feynman diagram are identical to the outgoing electrons.

How could I tell the particles were truly identical? I had learned from Phil Dauber (the physicist who had taught me that 95 percent confidence of time reversal violation is not high enough) that identical particles have waves that interfere with each other. In some directions the waves reinforce; in others they cancel. Such interference is seen among the particles coming out of a collision (part of their "final state interaction") and was readily observed in our data. Particles with different makeups don't interfere with each other. A pion doesn't interfere with an electron. One electron could interfere with another, but only if its hidden internal spin is identically oriented. Interference shows that particles are identical, even in all their possibly hidden interior structure—identical down to the full extent of quantum physics.

In my bubble chamber photos, I could see two identical pions, yet they disintegrated at different times. I still find that very strange. Two identical sticks of dynamite with identical fuses, if lit simultaneously, would explode simultaneously. My identical particles didn't. There had to be a difference between the two pions. They couldn't have identical wave functions. Yet their interference showed that they did.

For most radioactive atoms you can't be sure, even though the Freedman-Clauser experiment argues against hidden variables. My

---

short. I think the name was made extra long in hopes that most people would use the nickname; the name "Lawrence" had become associated with bombs. The Bevatron was named after the fact that it was the first machine to accelerate particles to a billion electron volts.

observation of different behavior for particles that were demonstrably identical removes that possible objection. Of course, I was not the first to do this. I learned the method from Dauber. All I am doing now is drawing your attention to a kind of observation that is well known in particle physics—one that is relevant to the discussion of physicalism and the extent to which the past determines the future.

I may not have free will, but these pions sure appear to have it.

No, I don't mean that they really do. It is a bit rash and anthropomorphic to say pions have free will. Rather, the example shows that the physicalist claim that the world is deterministic is falsified by physical observations. Identical particles do not behave identically. Therefore, given complete knowledge of the past, even with accuracy sufficient to defeat chaos, certain important aspects of the future (such as one that might affect the lifetime of a cat) cannot be predicted. The most powerful historical argument against free will, the argument that formed the success of classical physics, the argument that physics was deterministic, was itself the illusion.

Classical Free Will

What is free will? In the late 1800s, in the climactic period of classical physics, science was making enormous strides in explaining *everything*. The following quote is attributed to Lord Kelvin:

> There is nothing new to be discovered in physics now. All that remains is more and more precise measurement.... The future truths of physical science are to be looked for in the sixth place of decimals.

This statement (whether or not Kelvin actually said it) reflected what many scientists felt at the time. Everything: mechanics, gravity, thermodynamics, electricity and magnetism—everything—seemed to be falling into place. Soon even biological behavior would be reduced to bouncing particles and electrical signals. You had to be a scientific pessimist, maybe even a science denier, to think that free will would remain intact.

Philosophers analyzed free will at length and reached different conclusions. Schopenhauer presented his 1839 paper "On the Freedom of Will" not to a meeting of philosophers, but to the Royal Norwegian Society of Sciences. He argued that humans do not possess anything but the illusion of free will:

> You can do what you will, but in any given moment of your life you can will only one definite thing and absolutely nothing other than that one thing.

In *Beyond Good and Evil* (1886), Friedrich Nietzsche called free will a "folly" resulting from the extravagant pride of man, a "crass stupidity."

Immanuel Kant (1724–1804), known primarily for his philosophy, was also an outstanding scientist; he was the first to recognize that tides slow the rotation of the Earth, and he correctly hypothesized that our solar system formed from a primordial gaseous nebula. Kant had an excellent understanding of Newtonian physics and the possibility it presented that even life itself was deterministic. Yet he concluded that he had free will, despite the success of physics at the time, simply because (he argued) without free will there would be no difference between moral and immoral behavior. Since there is, he said, free will must exist.

That is rather conclusory reasoning, any lawyer would say today, yet I think Kant's statement has a deeper interpretation. He felt that he had nonphysics knowledge, true knowledge, of ethics and morality and virtue. Given his certainty of this knowledge, free will must indeed exist, for in the absence of choice those concepts could not have true meaning. But it would take advances in physics, particularly in understanding its quantum aspects, to see the true compatibility between physics and Kant's thoughts on free will.

A modern-day scientist-philosopher, Francis Crick, codiscoverer of the double-helix structure of DNA, disagreed. He propounded,

"You," your joys and sorrows, your memories and your ambitions, your sense of personal identity and free will, are in fact no more than the behavior of a vast assembly of nerve cells and their associated molecules.

Crick calls this his "astonishing hypothesis," even though, in my mind, he is simply recanting the numerous philosophers who based their conclusions either on an all-powerful and all-knowledgeable God, or on the remarkable success of classical physics. Strong opinions are not always backed by compelling reasons. To his credit, Crick called it a "hypothesis," not a conclusion that he could reach from science alone. Indeed, his conclusion, as well as that of Schopenhauer, was not falsifiable.

To repeat, the conclusion I reached in the section on my pion observations is not that pions have free will, or that people have free will, but that the philosophers' key assumption that the past completely determines the future is not supported in modern physics. Their arguments that free will does not exist were based on a false premise. We can't conclude that free will exists, but we can conclude that nothing in science rules it out.

Even though modern physics allows for free will, any exercise of free will must be compatible with the increase of entropy, the law that probable events are more likely than improbable ones. Entropy is an absolute constraint. Can free will overcome the tyranny of entropy?

Directing Entropy

Can nonphysics knowledge be used to affect the future, to move it in a direction that is less probable than it otherwise would have gone? The answer will prove important once we have a physical origin for the meaning of *now*, in determining why that moment is special for us.

I think the answer is obviously yes. Even if we can't decrease the entropy of the universe, we can manipulate its increase, direct its production to meet goals. We can exercise our free will by choosing which futures are accessible. We can choose where to place the teacup—in the center of the table or near the edge. We can choose whether or not to puncture the barrier between a container of gas and a container of vacuum. Entropy, then, will take us to the most probable state, but we get to choose the set of states. We are the conductor, and entropy is our orchestra.

Wood can rot—and increase its entropy—or we can strike a match and use that same wood to fire pottery and make a teacup or to push a piston to drive a tractor to build a city. Entropy still increases, but most of that increase can be dumped off into space in the form of discarded heat radiation. The local entropy, the entropy of the city, of our environment, of our civilization, can be induced to decrease. It decreases because we will it to do so. What I am saying is not new; it was described by Erwin Schrödinger in his 1944 book *What Is Life?*

Is the existence of free will a hypothesis that can be falsified? It is much harder to do experiments on humans than on pions, but at least we can consider whether a test could be done in principle, and what we mean by free will. Here is my attempt:

> If humans always follow the laws of probability, then free will does not exist. If humans regularly do highly improbable things, things that are not predictable based on external influences, then such behavior constitutes free will.

This statement contrasts directly with the claim of Schopenhauer, quoted earlier but worth repeating here: "You can do what you will, but in any given moment of your life you can will only one definite thing and absolutely nothing other than that one thing." Schopenhauer's claim was founded on a physicalist belief that was plausible back in the days of classical physics but is no longer credible. Even though he presented his paper to a science forum, Schopenhauer never proposed a method for falsifying his theory.

I can't offer a physicalist proof of the reality of free will. I argue simply

FREE WILL   283

that there is no valid proof of its nonexistence, not even a strong argument, and that nonphysics knowledge, along with the recognition that not all paths to increases in entropy are accessible, offers an alternative to the physicalist psychological-delusion explanation.

We Hold These Truths to Be Self-Evident

During the heyday of classical science, from Newton to Einstein, as physicists appeared to be showing that physics determined the future, there was perplexity in philosophy. As physics eroded the belief in an all-powerful and activist God, what was the source (if any) of virtue? With God receding, the Enlightenment in Europe was arguably an attempt to restore the foundation for human goodness that had once been dictated by a Supreme Being. What is the basis of ethical behavior? What sets the standards of morality, fairness, and justice? What about political rule? If the government wasn't established by God (divine right of kings), from where does it derive its power? What are the proper limits of that power?

During the Enlightenment, the divide between good and evil was accounted for in a pseudoscientific manner. Physics was the fad of that era. It set the standard for the use of reason to reach plausible explanation. Morality derived from reason. Virtue was justified by the value it created. In the 1700s, David Hume developed *empiricism*, something he called a "science of man," a method of accommodating moral responsibility even in a world that was deterministic. (It didn't really matter whether the determinism came from physics or from God.) Ethics was no longer based on abstract rules handed down by God, but on self-interest, and in the pleasure we obtain by helping others. Hume had deep insights that are still taken as valid, and he is considered to be a founder of the field of cognitive science.

The philosophy of the Enlightenment and post-Enlightenment cannot be summarized in a few books, let alone a few paragraphs. So forgive me for my overly brief treatment. But I think that period in philosophy was dominated by attempts to replace religion with concepts and ideals that would be just as powerful as the fading God, and could lead to principles for governing society. The philosophers struggled with logic, reason, and physics to try to explain why moral behavior continued to make sense, and why their new ideas of government were righteous. John Locke argued that human reason led to the recognition that humans are born with rights, rights later eloquently elaborated by Thomas Jefferson. Yet I feel that the role Locke attributed to reason was forced.

It is not *reason* that leads you to say that rights are self-evident, but empathy. Jean-Jacques Rousseau wrote of a fanciful primitive human society that was fundamentally peaceful. Thomas Hobbes made up a far-fetched story of the origins of government, explaining that it was a social contract between the rulers and the governed. Immanuel Kant, philosopher and physicist, attempted to develop a rationalist approach to morality. Jeremy Bentham elevated happiness as the measure for utility.

These thinkers talked about the ideal forms, the utopias. They created pseudoscientific equations, such as John Stuart Mill's goal of maximizing the greatest good for the greatest number—a concept that suggested an ability to compute the value of civilized conduct.* The Enlightenment philosophers were seeking a scientific justification for righteous behavior.

Now that I have trivialized the Enlightenment, where are we?

In my mind, the philosophers were on the right track. Their error was in thinking that the justification of their theories had to be based in a scientific structure, on reason and logic and science. Ultimately, the world is not deterministic, at least not the development of civilization. The future depends not just on the forces and motions of the past, not just on measurable physics, but on the perception of nonphysics reality and human action exercised through free will. This is a reality that cannot be readily quantified and cannot be reduced to reason and logic.

---

* Mill's concept was mathematically flawed. In general, you can't simultaneously maximize the result of two variables (good and number), only one.

Free Will and Entanglement

Could free will have a wave function basis? Yes, that's certainly possible. Let me engage in a little philosophical/physical speculation to illustrate this. I'll give an approach that is not a valid physics theory, because it is not falsifiable, but is interesting to ponder nevertheless.

Imagine that in addition to the physical world, there is a spiritual world. This is the world in which the soul exists; it is the realm in which empathy can operate and affect decisions. Imagine that the spiritual world is somehow entangled with the physical world. Action in the spiritual world can affect wave functions in the real world. The physical world can likewise inform and influence the spiritual one.

In ordinary entanglement, between two particles in the physical world, detection of one entangled particle affects the wave function of the other. Yet that entanglement is impossible to detect or measure if you are given physical access to only one particle. With both particles, you can see the correlation, but with only one, the behavior seems completely random.

When I try to understand my own soul, this picture makes some sense. There is a spiritual world separate from the real world. Wave functions from the two worlds are entangled, but since the spiritual world is not amenable to physical measurement, the entanglement can't be detected. Spirit can affect physical behavior—I can choose to build or smash a teacup; I can choose to make war or seek peace—through what we call free will.

This speculation is not falsifiable, but that doesn't mean it isn't true. As Gödel taught us, there are always truths that can't be tested.

Selfish Genes

Our employment of empathy and compassion, our sense of fairness and justice, could in principle have resulted from instincts developed during Darwinian evolution. This is the physicalist view, the belief that if something isn't measurable, it isn't real. It leads to a kind of relativism that makes some people uncomfortable. Virtue is no longer absolute, as it was in the days of deep religious conviction, but merely a result of our cultural evolution. We ought not to be arrogant about our moral beliefs, because they are temporary and culturally dependent, and in the future we may decide that our standards were badly distorted. After all, it wasn't that long ago that we imprisoned or even killed gays, and not long before that, slavery was widely accepted.

All our ethical goals can be interpreted as having Darwinian survival value, if not for the individual, then for the gene. Dawkins wrote eloquently about this theory in his fascinating book *The Selfish Gene.* Even altruism, Dawkins argues, is based on Darwinian evolution. We will readily sacrifice ourselves if it means that our genes, shared with our family and close relatives, our clan or cohort, will better survive. But although this theory is fascinating, is it right? That is much harder to determine.

Empathy does have positive survival value for the gene, but it also has negative survival value. Which one dominates? You don't want your soldiers to have too much empathy for the enemy that you wish them to kill. Empathy for outsiders is not obviously the result of a selfish gene. Dawkins would argue that the positive survival value dominates, but would he say that out of analysis or because it leads to his conclusion? Physicalists need to be careful about making an arbitrary assumption that the virtues are a result of evolution. That is not obvious, and it may not be true. It fits neatly into the belief that science can explain everything, but we know that science can't. Again, see Appendix 6 for statements by some prominent physicists who are not physicalists.

An alternative "explanation" for the origin of virtue is that it comes from our real, true, but unmeasurable nonphysics information. Compassion and empathy arise from the knowledge (belief? perception? guess?) that other people have deep inner essences, souls, just as you know you have yourself. It might be considered a religious revelation when you recognize (believe?) that other people are just as real as you are. The origin of love is empathy, not sex—although you may be influenced in your choice of sexual partners by your selfish genes. With empathy, you are led to feel (believe? know?) that the proper way to behave is to treat others exactly as you would have them treat you. Then most of the virtues can derive from that simple Golden Rule.

Richard Dawkins proudly proclaims himself to be an atheist—that is, not a theist. He claims to base his atheism on logic, but reasoning that ignores observation is not logical. His religion is physicalism.

Many atheists say they hold to no religion, and for some of them that might be right. But anyone who claims, "If it can't be measured, if it can't be quantified, it isn't real" is not without religion. Such people often (in my experience) believe that their approach is obvious, and therefore they call it logical. They hold their truths to be self-evident. It is worthwhile to recognize that not long ago, the fundamental tenets of Christianity were held to be self-evident, at least among most Europeans. Isaac Newton wrote religious tracts in which he described his literal belief in the Christian Bible.

As for understanding reality, it is time to recognize that physics is incomplete. Physicalism has been a powerful religion, very effective in advancing civilization by the focus it has given to physics, but not something that should be used to exclude truths that can't be quantified. There is reality beyond physics, beyond math, and ethicists and moralists should not abandon approaches solely because they have no scientific basis. Other disciplines need to pull back on their exaggerated physics envy and recognize that not all truths are founded in mathematical models.

PART V

NOW

24

The 4D Big Bang

*As the Big Bang creates new space, it also creates new time . . .
and that new time is the key to* now.

God, you can go crazy thinking about all this . . .
— *Sarah Connor, in*
The Terminator

Though the moments quickly die,
Greet them gaily as they fly.
— *Chorus of the young maidens,*
from Pirates of Penzance

The advances Einstein made in our understanding of time were monumental. Feynman found value in incorporating backward time travel. Since then, I believe progress in understanding time has been virtually nil.

In assembling a jigsaw puzzle, it is sometimes hard to find a missing piece, but the real impediment is the piece that has been put in the wrong place. The entropy explanation of the arrow of time was just such a misplaced piece. Civilization is built on the local decrease in entropy, not on its increase. Sure, a movie about a breaking teacup is a great example of entropy increase, and played backward it is completely implausible, but a movie of a teacup being manufactured would look just as wrong played backward.

The entropy of the Earth is decreasing as the core cools. Local entropy decrease is a characteristic of the spread of life and of civilization. Associating time with entropy *decrease* has the distinct advantage of being a theory in which it is the local change that is most important, not a change of some distant black hole. In fact, ultimately, the decrease in entropy is an essential part of what we call life: taking unorganized nutrients from soil and air and arranging them first into food (through plant production), then flesh (through eating and digestion), and then growth and learning. When the entropy of our bodies does finally begin to increase dramatically, we call that phenomenon death.

The Leading Edge of Time

Could the Big Bang itself be responsible for the flow of time? Yes, of course, say many theorists, but they feel obliged to include the entropy mechanism as a link between the expansion of the universe and the progression of time. The Big Bang put the early universe into a low-entropy state, giving room for entropy to increase. But why include entropy at all, when including it suggests results that are not observed, such as local correlations between the rate of time and entropy? Let's look at the Big Bang itself and see how it could be directly responsible for the flow of time, for the meaning of *now*, with no need to invoke entropy as a crutch.

In the modern cosmological picture, the Lemaître approach, the galaxies don't move—at least, not significantly; except for a small "proper motion" (such as our local acceleration toward Andromeda), they rest at fixed coordinates. The Hubble expansion represents not the movement of galaxies, but the creation of new space. This creation of new space is not mysterious; general relativity gave space flexibility and stretchability. Space can expand readily, but when it does, the future of that expansion is governed by the equation of general relativity, the equation that says the geometry of space is determined by its energy-mass content, the equation that, in its most elegant form, looks deceptively simple: $G = kT$.

Is the Big Bang an explosion of 3D space? Yes—but a more reasonable assumption, one closer to the spirit of space-time unification, is that the Big Bang is an explosion of 4D *space-time*. Just as space is being generated by the Hubble expansion, so time is being created. The continuous and ongoing creation of new time sets both the arrow of time and its pace. Every moment, the universe gets a little bigger, and there is a little more time, and it is this leading edge of time that we refer to as *now*.

Although many people find the continual creation of space to be counterintuitive, the continual creation of time fits right into our sense of reality. It is exactly what we experience. Every moment, new time appears. New time is being created right *now*.

The flow of time is not set by the entropy of the universe, but by the Big Bang itself. The future does not yet exist (despite its inclusion in standard space-time diagrams); it is being created. *Now* is at the boundary, the shock front, the new time that is coming from nothing, the leading edge of time.

Are All *Nows* Simultaneous?

Is your *now* the same as my *now*? Let's first consider this question in the usual reference frame of cosmology, the reference frame described by George Lemaître. He had all the galaxies at rest, and the space between them expanding. Every galaxy can be thought of as having a clock on it. According to the cosmological principle (which is built into Lemaître's model), all galaxies everywhere look the same; they all have experienced the same time since the Big Bang, and all the clocks will read the same. That means they will all experience *now* simultaneously.

But as with special relativity, the concept of simultaneity can depend on reference frame. Consider a frame with the Milky Way at the center. In this frame, all the galaxies are moving away from us, and time is dilated in those galaxies; it is running slower and the *nows* are no longer synchronous. In this frame, the time passed since the Big Bang is larger for us than it is on other galaxies. The concept of *now* is no longer simultaneous across the universe. Our *now* occurs first.

As with special relativity, this behavior of simultaneity is not a contradiction; it is a *feature* of general relativity.

The Perception of *Now*

Why do you feel that you exist in the present? Actually, you exist in the past too; you know that quite well. You exist backward in time right up to the moment you were born (or conceived, depending on your definition of life). Your focus on the present comes largely from the fact that, unlike the past, it is subject to your free will. According to physics, as we currently understand it, the past does not totally determine the future; at least some random element originates in quantum physics. The presence of such a random element means that physics is *incomplete*, that the future is not determined uniquely by the past, that nonphysics realities could play a role in determining what is going to happen. The fact that physics is incomplete leaves open the possibility that we can also affect the future by use of free will.

I can't prove that free will exists, but when physics includes quantum uncertainty, it no longer can deny the possible existence of free will. If you have free will, then you can exercise your nonphysics knowledge to open or close the possible paths of entropy increase, and thereby affect what is happening and what is going to happen. You can smash a teacup or build a new one; probability and entropy have nothing to do with your decision. To quote John Dryden, *What has been has been. Not heaven itself upon the past has power.* And—bad news for science fiction fans—neither do you. No loop through a wormhole can change that.

Through its pervasive use of the space-time diagram, in both research and teaching, physics effectively avoided the issue of the flow of time. The time axis is treated (mostly) as another space axis; its special feature of the progression of time is completely missing. *Now* is just another spot on this axis, as if the future already exists but has just not yet been experienced. Time travel would consist of altering that *now*, moving it forward or backward along the axis. But *now* is not movable. *Now* is the leading edge of the 4D Big Bang. *Now* is the moment that has just been created. The time axis for a true space-time diagram does not extend to infinity. Time stops at *now*.

Can the future affect the present? What about positrons, electrons moving backward in time, coming from the future to involve themselves in present interactions? Yes, that is the current approach of physics, a physics that ignores *now*, that is based on the infinite space-time diagram. Does that current approach, so successful in calculating the electron magnetic strength to better than ten decimal places, mean that all of its assumptions are valid? Many physicists think yes, at least until we have an alternative.

Perhaps a kind of uncertainty principle is at work. The future can affect the present only to the extent that part of the future is already determined, and therefore is inherent in the present. Hawking argued for that; he wrote that backward time travel is possible only on the microscopic scale. Presumably he would not accept the positron photographed by Anderson as a particle moving backward in time.

I will argue, however, that the distant future does not exist, not yet, not in the sense that the present and the past exist. The past has been determined; what has been has been; the future is still not there because we know it is not predictable, not by the current laws of physics, which are unable to predict even when a radioactive atom will decay. Religious determinists thought the future was already set, by the perfection and foresight of their all-knowing God. And for a while, we thought we didn't need such a God to have determinism; we thought physics by itself could do that. Now we know better.

Dirac's equation predicted the existence of antimatter, and Feynman eliminated an absurdity in Dirac's interpretation, the infinite sea of filled negative-energy states, by recognizing that the antimatter solutions could be construed as negative energy particles moving backward in time—giving them, effectively, positive energy. That's history. Feynman recognized that backward negative-energy states are indistinguishable from forward positive-energy states. But let's not take the backward-in-time interpretation too seriously. Positrons do exist; they have positive energy, and they do indeed move forward in time, not backward.

What has been has been. If Dirac's equations predicted the positron's existence through a series of convoluted interpretations, fine. Here's a historical analogy. Niels Bohr came up with the first model that correctly accounted for the spectrum of hydrogen; in 1913 that model gave an enormous boost to the fledgling field of quantum physics. We now know that Bohr's theory was wrong; it makes definite predictions (for example, about the angular momentum of the lowest-energy orbiting electron) that are incorrect and falsify the theory. No matter. Thirteen years later, both Heisenberg and Schrödinger came up with better theories, inspired in large part by Bohr, and these theories gave exactly the same hydrogen spectrum but did not make the incorrect predictions.

We still honor Bohr as one of the founders of quantum physics. We still teach Bohr's model to new students; it is a simple and compelling way to introduce the study of quantum behavior. (Very few professors point out that the model makes falsified predictions; they don't want the students to know that the intuitive and simple Bohr model is wrong—at least, not until their education in physics goes further.) We will someday feel the same way about Dirac and Feynman and their far-fetched theories of antimatter.

Falsifying the Cosmological Origin of Time, Part I

Is the cosmological origin of the arrow of time—including the creation of new time by the Big Bang, the flow of time, and the meaning of *now*—a falsifiable theory? One possible way to test it uses the discovery that the expansion of the universe is accelerating, that the universe is growing at an ever-increasing rate. Time is linked to space; it is the fourth dimension of space-time, so it is natural to expect that the rate of time is also accelerating. That means that clocks today are running faster than they were yesterday, that they exhibit a *cosmological time acceleration*. Could this acceleration of time be detected and measured?

In principle, the answer is *yes*, variability in the rate of universal time can be detected by looking a distant clocks.

Recall that a small difference in clock rates was detected in the Pound-Rebka falling-gamma-ray experiment, in which the time dilation due to gravity was first observed. It was also seen in the Hafele-Keating airplane experiment, in which high-altitude clocks were observed to run faster than clocks left on the ground—and slower from the velocity effect. The difference is seen every day with GPS, which also has to correct for these time effects. Gravity's effect on time is observed when we measure spectral lines on the surface of white dwarf stars; they show a frequency shift from the time dilation because the strong gravitational fields slow time on their surfaces.

In principle, any one of these experiments could also detect accelerating time. The signals are emitted at one time, pass through space, and are received later. Most of the observed effect comes from the gravitational potential and Doppler shift, but a slight excess would come from cosmological time acceleration. The effect would be independent of direction; it would always be a *redshift*; that is, the observed rate from the past would always be slower than the rate of the present clock. The Pound-Rebka experiment showed increasing frequency for falling gamma rays, and would (presumably) show decreasing frequency for rising ones; the cosmological time acceleration would have decreasing frequency for both.

We might also look for an anomalous redshift for distant galaxies. The galaxies for which the acceleration is most precisely measured emitted their light about 8 billion years ago. Their different velocity, compared to the Hubble expansion velocity, has been observed to be about 4 percent. These galaxies are 8 billion light-years away and are receding (distance increasing) at 40 percent the speed of light. The part of this velocity that is due to time acceleration is about 2 percent of lightspeed.

Of course, all distant galaxies already show a redshift, but we attribute that to the expansion of space, the fact that the distance to the galaxy is rapidly increasing. That's Hubble's law. How can we distinguish the redshift that is due to the expansion from the redshift that is due to the cosmological time dilation? One way might be to make a separate measurement of the distance change, one that does not depend on the velocity redshift. If we knew the rate of change of distance, then we would know how much of the redshift was due to expansion, and how much was left over to attribute to the cosmological time dilation.

Before we look for a way to do this (that is, a way that might be completed within my lifetime), let's consider whether the experiment could be done *in principle*—that is, if we had unlimited resources and unlimited patience. Suppose we had a billion years to do the experiment. Couldn't we just see the rate at which the galaxy was receding, without depending on the velocity redshift? We could try to find a "standard ruler" in the galaxy, perhaps the size of a known kind of star, and observe how the apparent size of that ruler changes with time—thus getting an independent recessional velocity estimate. Or maybe we could detect light (microwave radiation?) that reflects off the galaxy. The goal is to separate redshift that depends on recessional velocity from redshift that also depends on intrinsic time dilation.

There is a catch. Our current notion of distance is dependent on our measure of time. We currently *define* the length of a meter to be the distance that light travels in vacuum in 1/299,792,458 of a second. This definition means that light, or any truly massless particle, travels through empty space at exactly 299,792,458 meters per second. So no experimental measurement can ever determine the speed of light more precisely! The reason for defining length in this way is not that we're lazy; it turns out to be very difficult to come up with a good definition for the meter, and this is the best we have found. It replaced the old method of having the standard depend on a meter stick stored in a vault in Paris. But if the clock is slower in that distant galaxy (compared to our clock), then the ruler, the standard meter stick on a planet in that galaxy, will be larger, since light will travel farther each second. That means the measure of distance, taken from the standard size, will be different. Cosmological time dilation could be confused with a change in the rate of expansion.

In fact, a look at the equations of the Lemaître model suggests that this problem may be intractable, at least to the extent that the cosmological principle (a perfectly uniform universe) is exact. There may be no way to distinguish the expansion of space from a dilation of time. Of course, the universe is not completely uniform; the cosmological principle is just an approximation that enables us to do calculations and find the solution in terms of a simple (to physicists) mathematical expression. Perhaps we can take advantage of the nonuniformity of space to detect the acceleration of time. Perhaps this acceleration can be detected locally; the Pound-Rebka experiment (dropping gamma rays from a tower) managed to see a frequency shift that was only one part in a million-billionth (10^{-15}). I have no practical suggestion, as of today. I take heart in the fact that when Dirac proposed his positron, he believed there was no way to detect it for the foreseeable future.

Falsifying the Cosmological Origin of Time, Part II

Another possible way to falsify the cosmological origin of time depends on the reality of inflation theory, the idea that in the first millionth of a second, the universe expanded at a speed greatly exceeding that of light. This acceleration period was a precursor to our current acceleration period, and if the 4D explanation of time is correct, then not just space, but time too, should have inflated. Could we observe the first millionth of a second of the Big Bang?

Remarkably, the answer is maybe. Currently, the earliest probe we have is the cosmic microwave pattern, which examines the time a half-million years after the beginning. But a potential signal was emitted earlier, in the first millionth of a second: gravitational radiation. There is hope that we may soon be able to detect these primordial gravity waves, and they have the advantage of probing much closer to the moment of creation, possibly within the time we need to observe inflation. The way to see the gravity waves is to look at the pattern that they induced on the cosmic microwave radiation, particularly on its polarization.

For a while, some physicists thought just such a pattern had been observed. An initial report of the discovery of such gravity waves in March 2014 from a project called BICEP2, short for "Background Imaging of Cosmic Extragalactic Polarization 2." This project measures the microwaves from a station at the South Pole, where the extreme cold removes the atmospheric water vapor that otherwise interferes with ground measurements. Unfortunately, the result proved to be a false alarm, likely a consequence of interference from emissions of cosmic dust.

New and more sensitive measurements are planned, and there is realistic hope that we will soon be seeing gravitational waves from the extremely early universe, from the period of inflation, and it may be possible to distinguish a space inflation from one that involves both space and time.

The Future of Physics

Sometimes I wish that Plato had been right, that these issues could all be settled with dialogue and pure thought, that the mind is the ultimate arbiter of truth. But the history of physics argues that Plato was wrong. We need to keep in touch with the physical world, just as Antaeus had to keep his feet on the ground.

Quantum entanglement is here to stay. Spooky action at a distance is no longer a speculation, but an experimental result demonstrated by Freedman and Clauser and by many subsequent experiments. Even though we can't transfer material or information faster than lightspeed, the instantaneous wave function collapse is an annoying problem suggesting that another approach might yield new insights. I harbor hope that someone will reformulate quantum physics without the need for amplitudes. Theorist Geoffrey Chew at Berkeley tried to do that when I was a student with an approach he called "S-matrix theory," but although it led in some significant ways to the modern standard model, it ultimately failed in its goal to eliminate quantum amplitudes and wave functions. Meanwhile, efforts to find whole new approaches have been put on hold because of the enormous success of the "standard model" of elementary particles. The standard model is the best theory ever produced in the history of physics, in terms of its ability to make precise predictions that are then verified experimentally.

So why change our quantum physics theory if it works so well? Despite the standard model's success, I think the reformulation will come. When it does, the amplitudes will no longer collapse with super-lightspeed, and (I guess) positrons will be neither holes in an infinite sea of negative-energy particles nor electrons moving backward in time. That was a handy way to look at them in the context of space-time diagrams, in which the flow of time, the progression of time, is completely absent.

The great step in quantum physics that is desperately needed is an understanding of measurement. Few physicists believe that human consciousness is required to make a measurement. Schrödinger made that case convincingly with his cat. But what is a measurement? Roger Penrose has argued that there is a micromechanism, a part of nature that makes many measurements. The quantum state that led to the structure we see in the Big Bang did not have to wait until Penzias and Wilson discovered the cosmic microwave radiation, and the Milky Way was not at rest in the universe until that moment when my team deduced its velocity. (Was it when the apparatus measured the anisotropy, or was it when I looked at the data?) The moon was there even before Einstein looked at it. Something natural had already made the wave function, a superposition of an infinite number of possible universes, collapse long before humans (or animals) appeared.

Advances in technology have made experimental research on the theory of measurement far more tractable. You no longer need atomic beams of calcium to create entangled photons; you can make them by shining a laser beam on a special crystal, such as BBO (beta barium borate) or KTP (potassium titanyl phosphate). As a result, experiments probing quantum measurement have been making remarkable strides.

One of the more intriguing results has been the study of "delayed choice," in which measurements are collected for all orientations of

Figure 24.1. Jeremy and Pierce ponder the flow of time. From the *Zits* comic.

polarization and only afterward are the data analyzed. Such experiments probe the question of whether human decision must be involved in making a measurement, and they indicate that the answer is no. OK, that's no surprise, but the real breakthrough will come when there is a surprise, as there was with the Michelson-Morley experiment.

The new laser methods have made it possible to test entanglement at far greater distances than those attempted by Freedman and Clauser. On October 22, 2015, a front-page headline in the *New York Times* proclaimed, "Sorry, Einstein, but 'Spooky Action' Seems Real." A group at Delft University of Technology in the Netherlands had verified superluminal entanglement effects between two electrons separated across campus, a distance of nearly a mile. Once again the Copenhagen interpretation, with its faster-than-light action, could proclaim victory.

The observation in 2015 of a gravitational wave by LIGO (see www.ligo .caltech.edu) suggests a third test of the *now* theory of time creation. As two black holes collapse into each other, new time should be created locally, observable as an increasing delay between the predicted and observed signal. The one wave seen so far is insufficiently precise to test this prediction, but if many events are seen, or a closer one with a stronger signal, then the presence or absence of this lag could confirm or falsify the *now* theory.

25

The Meaning of *Now*

The puzzle pieces are all in place. What does the picture look like?

Not heaven itself upon the past has power;
but what has been has been, and I have had my hour.
— *John Dryden (1685)*

E instein took the first great step in the quest for the meaning of *now* when he realized that space and time are flexible. Lemaître applied Einstein's equations to the universe at large and developed a remarkable model in which the space of the universe is expanding. A few years later, when Hubble discovered that the universe was indeed expanding, the Lemaître model—independently developed by Friedman, Robertson, and Walker—became the standard model, the way all cosmologists currently interpret the Big Bang.

The puzzle began to fit together, but there were several impediments—jigsaw pieces jammed into the wrong places. One of these was Eddington's attribution of the arrow of time to the increase of entropy. In 1928, when he proposed this, Eddington didn't know that the dominant reservoirs of entropy were the unchanging microwave radiation and the remote surfaces of black holes and the distant edge of the observable universe. As Schrödinger pointed out, civilization depends on local decrease in entropy, but that local decrease, in the Eddington approach, plays no role in the entropy arrow of time.

Another wrongly placed piece was the misinterpretation of the space-time diagram. It shows no flow, no moment of *now*, so it offers a ready excuse to avoid these issues. Some theorists even interpret the absence as an indication that they are meaningless concepts, illusions that play no role in reality. The mistake of this vision is in interpreting a computational tool as a deep truth. It is fundamentally the error of physicalism: *if it isn't quantifiable, it isn't real.* In fact, it is based on an extreme fundamentalist version of physicalism: *if it isn't in our current theories, it isn't real.*

The third wrong piece was related to another aspect of physicalism: the assumption, made by Einstein and others, that the past can be, must be, able to determine the future completely. The driving philosophy was the principle that physics should be *complete*. If quantum physics did not allow the time of a radioactive decay to be predicted, then that was a fault of quantum physics that needed to be corrected. This assumption used to deny free will, the ability to choose.

Pull out the misplaced pieces, some of which are not even part of the puzzle, and the rest fits together in a natural way. As space expands, so does time. The elements of time in which quantum physics has already operated, through a mysterious measurement process that we don't yet understand, are what we call the past. We live in the past just as we live in the present, but we can't change the past. *Now* is that special moment in time that has just been created in the expansion of the 4D universe, as part of the continuing 4D Big Bang. By the *flow* of time, we mean the continual addition of new moments, moments that give us the sense that time moves forward, in the continual creation of new *nows*.

Now is the only moment when we can exercise influence, the only moment when we can direct the increase in entropy away from ourselves so that we can orchestrate local entropy to decrease. Such local decrease is the source of expanded life and civilization. To direct entropy in that way, we must have free will—a capability that physicalists call an illusion, even though current quantum physics theory has similar behavior built into its essence.

The existence of free will could be falsified if we found faster-than-light tachyons, those particles that in some reference frame would imply that the consequences preceded their causes. Perhaps we will discover, by studying entanglement as a function of direction (parallel to and perpendicular to the Milky Way proper motion), that there is a special reference frame for causality. The top candidate is the Lemaître frame, the only frame in which all the *nows* across the universe are being created simultaneously. If that proves true, then we must modify relativity.

It is conceivable that the uncertainty principle will someday be shown to be wrong, to have been only an uncertainty in our current physics theory, not present in the more complete version that will replace it. But the Freedman-Clauser experiment, showing the reality of entanglement, suggests that spooky action at a distance will not go away. It is not any one physics theory that is incomplete; it is physics itself. That is evident from the fact that physics alone would not be able to discover, let alone prove, that $\sqrt{2}$ is irrational. It is evident from the fact that clear concepts that are easy to understand, concepts that are at the core of our experience of reality—such as *what does the color blue look like?*—lie outside the purview of physics.

Attempts to attribute all altruistic behavior to the instinct for survival, of the fittest organism or of the fittest gene, should be recognized as hypotheses, speculative attempts to give a pseudoscientific explanation for virtue, based on the physicalist dogma that everything can be explained through science. That is an unproven supposition based on anecdotal evidence, not in the class of Darwinian evolution (which has an enormity of data behind it), not a conclusion based on compelling scientific evidence, such as relativity and quantum theory. Physicalism can serve usefully as a working principle in the physics profession, just as a belief in capitalism can help you run an economy, but you don't want to drift into the mistake of thinking that either physicalism or capitalism, because of success in advancing our standard of living or in winning wars, therefore represents the entirety of truth.

Rejection of physicalism brings us to ponder the source of empathy.

Do we love our children and grandchildren simply because they carry the same genes that we carry, or is it something more profound, something about not just recognizing but actually perceiving the reality of the souls of those who are close to us? Ideas of ethics, morality, virtue, fairness, and compassion, the difference between good and evil, may all be linked to the fundamental empathy perception—something that goes beyond the gene and beyond physics.

Free will is the ability to use nonphysics knowledge to make decisions. Free will needs do no more than choose among the accessible futures. It doesn't stop the increase of entropy, but it can exercise control over accessible states, and that gives entropy direction. Free will can be used to break a teacup or to make a new one. It can be used to start a war, or to seek peace.

OFTEN THE MOST DIFFICULT challenge is in asking the right questions. It is hard to know where the next physics revelations will appear. Einstein showed us that time is a suitable subject for physical study. I think he was unable to address the meaning of *now* for the simple reason that he refused to accept the notion that physics was incomplete.

We may not soon understand the interplay between relativity and quantum physics or the meaning of measurement, but the issues are worthy of further attack. Advance is unlikely, in my mind, to require complex math or arcane philosophy. Whoever cracks open these problems will likely do so with some very simple examples, maybe using nothing more than algebra, perhaps referring to the little hand of a watch and where it points. It may happen when some simple experiment gives a result that is unexpected. When the next breakthrough happens, I predict it will take a regression to childhood, a way of looking at reality that focuses on something in physics that we don't even realize we're assuming is true, and turns it on its head. Who will this new Einstein be? You?

APPENDICES

APPENDIX 1.
THE MATH OF RELATIVITY

This appendix is meant for those who would like to see the algebra and the calculations behind the relativity results discussed in the text.

In special relativity, an event is labeled by a position x and a time t. To keep everything simple, let's set the other position coordinates, y and z, equal to zero. We'll label the position and time of the events in the second coordinate system, moving at velocity v, with capitals: X and T. Einstein determined that the correct relationships of x, t, X, and T were given by the Lorentz transformation:

$$X = \gamma(x - vt)$$

$$T = \gamma(t - xv/c^2)$$

where c is the speed of light, and the time dilation factor gamma is represented by the Greek letter γ and is given by $\gamma = 1/\sqrt{1 - \beta^2}$, where the Greek letter β (beta) is the lightspeed (velocity in terms of the speed of light: $\beta = v/c$). There is an implicit convention in these equations that the special event $(0, 0)$ has the same coordinates in both frames.

Hendrik Lorentz was the first to write out these equations; he showed that Maxwell's equations of electromagnetism satisfied them. But it was Einstein who recognized that they represent true changes in the behavior of space and time, and who then derived the new equations of physics using them. Maxwell's equations didn't have to change, but Newton's did, and Einstein concluded, among other things, that the mass of moving objects increases (I'm talking here about kinetic mass, given by γm), and that $E = mc^2$.

A remarkable feature of the Lorentz transformation equations is that solving them for x and t produces equations that look the same, except for the sign of the velocity. (The algebra is a little tricky, and you have to use the definition of γ provided above, but give it a try.) The answer is

$$x = \gamma(X + vT)$$

$$t = \gamma(T + Xv/c^2)$$

Compared to the previous equations, the sign change (from – to +) is what you would expect because with respect to the second frame, the first frame is moving at velocity $-v$. Nevertheless, the fact that the equation has the same form is amazing to me. I wouldn't have guessed that would happen. The fact that it does is part of the miracle of relativity, that all reference frames are equally valid for writing the equations of physics.

Time Dilation

Now let's look at time dilation. We'll use the same terminology as in the twin-paradox example discussed in Chapter 4. Recall that Mary travels to a distant star while John stays at home. We'll call the first frame John's frame, and the second one, moving with relative velocity v, Mary's frame. (These are their *proper* frames.) Consider two events: Mary's birthday party 1 and Mary's birthday party 2. We'll label the position and time for these two parties, in John's frame, as x_1, t_1 and x_2, t_2. The positions and times in Mary's frame will be X_1, T_1 and X_2, T_2.

Now let's put these values into the Lorentz equations. We'll use the second set:

$$t_2 = \gamma (T_2 + X_2 v/c^2)$$

$$t_1 = \gamma (T_1 + X_1 v/c^2)$$

By subtracting these two equations, we get

$$t_2 - t_1 = \gamma [T_2 - T_1 + (X_2 - X_1)v/c^2]$$

Mary's age as measured in Mary's frame is $T_2 - T_1$. In this frame, Mary is not moving, so $X_2 = X_1$, and therefore $X_2 - X_1 = 0$. So the equation simplifies to

$$t_2 - t_1 = \gamma (T_2 - T_1)$$

We can make this equation look simpler by using the notation that $\Delta t = t_2 - t_1$, and $\Delta T = T_2 - T_1$. (Δ is the Greek capital letter "delta" and is often used to represent differences. Out loud, we read Δt as "delta t.") With this notation, the equation becomes

$$\Delta t = \gamma \Delta T$$

This is time dilation. The time between the two events in John's frame is bigger than the time between the same events in Mary's frame, by the factor γ. In the twin-paradox example described in Chapter 4, γ was 2, so it takes Mary 16 years (in John's frame) to age by 8 years.

Length Contraction

Now let's look at length contraction. To measure distance in any frame, we note the locations for simultaneous times and subtract them. The distance between two simultaneous events ($t_2 = t_1$) in John's proper frame is $x_2 - x_1$. We apply the first set of Lorentz equations to the two events:

$$X_2 = \gamma(x_2 - vt_2)$$

$$X_1 = \gamma(x_1 - vt_1)$$

Subtracting these equations gives

$$X_2 - X_1 = \gamma[x_2 - x_1 - v(t_2 - t_1)]$$

Since for this example the two events are simultaneous in John's frame, we have $t_2 = t_1$, so the term $(t_2 - t_1) = 0$. Putting this in, the equation simplifies to

$$X_2 - X_1 = \gamma(x_2 - x_1)$$

The distance between the two events, in John's proper frame, is $x_2 - x_1$; let's call that Δx. The length of the object in Mary's proper frame (in which it is stationary) is $X_2 - X_1$; let's call that ΔX. Then we have the equation

$$\Delta x = \Delta X/\gamma$$

That's the length contraction equation. If the proper length of an object is ΔX, then, measured in a different frame, it will be shorter by a factor of $1/\gamma$. (Note that γ is always greater than 1.)

Simultaneity

The time difference between two events is $t_2 - t_1 = \Delta t$. In a different frame, the events occur at times T_2 and T_1, and the interval in that frame will be $T_2 - T_1 = \Delta T$. We'll also call the difference in positions of the two events in John's frame Δx, and the distance between the two events in Mary's frame ΔX. Using the first Lorentz transformation for time, we get

$$T_2 = \gamma(t_2 - x_2 v/c^2)$$

$$T_1 = \gamma(t_1 - x_1 v/c^2)$$

We subtract these and substitute Δt, ΔT, and Δx to get

$$\Delta T = \gamma(\Delta t - \Delta x v/c^2)$$

In the special case when the two events are simultaneous in John's reference frame (that is, when $\Delta t = 0$) the equation simplifies to this:

$$\Delta T = -\gamma \Delta x v/c^2$$

The remarkable result is that ΔT is not necessarily zero; that is, the events are not necessarily simultaneous in Mary's proper frame, even though they were simultaneous in John's proper frame. If I designate the distance between the two events $\Delta x = -D$ (the sign could be positive or negative, depending on the locations x_1 and x_2), then the equation becomes

$$\Delta T = \gamma D v/c^2$$

If neither v nor D are zero, then ΔT is not zero, and that means the two events are not simultaneous in Mary's frame. This is the "time jump" that takes place at a distant event when switching from one reference frame to another. No jump takes place when $D = 0$, that is, when the two events are at the same location (such as when John and Mary reunite). ΔT could be positive or negative, depending on the signs of D and v.

Velocities and the Speed of Light

I'll show here why the speed of light is the same in all frames of reference.

If an object is moving, we can call x_1 its position at time t_1, and x_2 its position at time t_2. Think of these as two events. The object's velocity is $v = (x_2 - x_1)/(t_2 - t_1) = \Delta x/\Delta t$. In another frame of reference, its velocity is $V = (X_2 - X_1)/(T_2 - T_1) = \Delta X/\Delta T$. We can use the Lorentz transformation to compare the two. Let's use the symbol u for the relative velocity of the two frames, so that we can use v and V for the velocity of the object in each of the two different frames. Write down the transformations for two events and subtract them:

$$\Delta X = X_2 - X_1 = \gamma[(x_2 - x_1) - u(t_2 - t_1)] = \gamma\,[\Delta x - u\Delta t]$$

$$\Delta T = T_2 - T_1 = \gamma\,[(t_2 - t_1) - u(x_2 - x_1)/c^2] = \gamma[\Delta t - u\Delta x/c^2]$$

Now divide the two equations to cancel the γ's:

$$V = \frac{\Delta X}{\Delta T} = \frac{\Delta x - u\Delta t}{\Delta t - \Delta x \frac{u}{c^2}} = \frac{\frac{\Delta x}{\Delta t} - u}{1 - \frac{\Delta x\,u}{\Delta t\,c^2}} = \frac{v - u}{1 - \frac{vu}{c^2}}$$

That's the equation for the velocity transformation. It gives the velocity V in the second frame in terms of v, the velocity in the first frame.

Suppose $v = c$; that is, an object (for example, a photon) is moving at lightspeed in the first frame. In the second frame its velocity is

$$V = \frac{v - u}{1 - \frac{vu}{c^2}} = \frac{c - u}{1 - \frac{u}{c}} = \frac{c\left(1 - \frac{u}{c}\right)}{1 - \frac{u}{c}} = c$$

regardless of u, the relative velocity of the two frames. If $v = c$, then $V = c$. Objects moving at the speed of light in one frame are moving at the speed of light in all frames. Try plugging in $v = -c$ and see what you get. Surprised?

A similar derivation shows that c doesn't change even if the direction of the light is arbitrary.*

---

* For arbitrary direction of light, you need to employ Einstein's additional transformation equations $Y = y$ and $Z = z$. Begin with $v_x^2 + v_y^2 + v_z^2 = c^2$ and calculate V_x, V_y, and V_z. You'll find that $V_x^2 + V_y^2 + V_z^2 = c^2$, but that the direction of the light changes. The directional change is called the *aberration of starlight* and is readily observed as a shift in the apparent direction of a star viewed from the moving Earth.

This result accounts for the failure of the 1887 Michelson-Morley experiment to detect a different velocity of light in two directions: the first parallel to the Earth's motion; the second, perpendicular to it.

Flipping Time

Something very interesting happens if two separated events are close in time. We'll use the difference equation (from the discussion of simultaneity above):

$$\Delta T = \gamma (\Delta t - v\Delta x/c^2)$$
$$= \gamma \Delta t[1 - (\Delta x/\Delta t)(v/c^2)]$$

Define $\Delta x/\Delta t = V_E$. This is the pseudovelocity that "connects" two events. It doesn't mean that anything actually travels between the two; it is the velocity that something would have to travel to be present at both events. Could V_E be greater than c? Yes, of course. Any two separated events that occur simultaneously have infinite V_E. This is not a physical velocity. Using this new terminology for V_E, we can write

$$\Delta T = \gamma \Delta t(1 - V_E v/c^2)$$

Let's take the example that Δt is positive. The equation shows that ΔT could be negative. All it takes is for the negative term inside the parentheses to be greater than 1. That means the order in which the events occur can be opposite in the new frame. This result has all sorts of implications for causality.

For $V_E v/c^2$ to be greater than 1, V_E/c must be greater than c/v. Remember, v is the velocity connecting the two reference frames; it must always be less than c. That means c/v will always be greater than 1. This equation means that if V_E/c is greater than c/v (thus also making it greater than 1), then the order of events in the two frames is reversed. Note again that there is no restriction on the size of V_E, since it is a pseudovelocity, the velocity needed to "connect" two events, and for two widely separated events that occur simultaneously V_E would be infinite.

Math of the Pole-in-the-Barn Paradox

Refer to the diagram in Chapter 4 (Figure 4.1, page 53). In the frame of the barn, the pole enters the door and continues until it reaches the back wall. Let's define $t_1 = 0$ as the moment when the front end of the pole hits the back wall and set up our coordinates so that $x_1 = 0$ that location. Because of the Lorentz contraction, in the frame of the barn the back of the pole enters the door simultaneously, at $t_2 = 0$, at location $x_2 = -20$ feet.

Now let's calculate what happens in the proper frame of the pole. The front of the pole hits the wall of the barn at time T_1, given by the Lorentz transformation equation:

$$T_1 = \gamma(t_1 - x_1 v/c^2)$$
$$= 2(0 - 0v/c^2) = 0$$

The back of the pole enters the door at

$$T_2 = \gamma(t_2 - x_2 v/c^2)$$
$$= 2(0 + 20v/c^2)$$

Calculating v/c from $\gamma = 2$, we get $\beta = v/c = 0.866$. So,

$$T_2 = 2(0 + 17.32/c) = 34.6/c$$

Using the speed of light $c = 10^9$ feet per second, we find that the pole will enter the door in $T_2 = 34.6/10^9$ seconds $= 34.6 \times 10^{-9}$ second. So when the front of the pole hits the wall, the back of the pole has not yet entered the door. It enters 34.6 nanoseconds (billionths of a second) later.

Let's calculate, in the pole frame, where the back of the pole is when the front of the pole hits the wall. We use the equation

$$x_2 = \gamma(X_2 + vT_2)$$

Solving for X_2 and putting in $0.866c$ for v, -20 feet for x_2, and $34.6/c$ for T_2, we get

$$X_2 = x_2/\gamma - vT_2$$
$$= -20/2 - 30 = -40 \text{ feet}$$

This answer agrees with our expectation. In the pole frame, when the front of the pole hits the wall, the back of the pole is at -40 feet. It is 40 feet from the back of the barn—consistent with the fact that in this frame it is 40 feet long.

The solution to the paradox is that the two ends of the pole are in the barn simultaneously in the barn proper frame, but in the pole proper frame, although they both get inside the barn, the entry of the back of the pole is not simultaneous with the crashing of the front of the pole into the wall. Once the pole is inside, if the motion of the pole is suddenly stopped (both ends are stopped simultaneously in the barn frame), it will lose its space contraction and suddenly expand to its full 40-foot length, crashing through either or both of the walls.

Math of the Twin Paradox

Because Mary's time dilation is $\gamma = 2$, we can calculate that her light-speed is $\beta = 0.866$. The twin-paradox example has several important reference frames: John's (what we'll call the *Earth frame*), Mary's *outgoing frame* (her proper frame on her way out, moving at velocity $v = 0.866c$), and Mary's *return frame* (her proper frame on her way back, moving at velocity of $-0.866c$). Mary's *proper frame* is a combination of these, as she accelerates from one Lorentz frame to another.

In the Earth frame, we can calculate the distance to the star from the fact that Mary travels at $0.866c$ and takes 8 years to get there; the distance is $0.866 \times c \times 8 = 6.92c$, or 6.92 light-years. In Mary's outgoing and return frames, the distance is $6.92c$ divided by the Lorentz contraction factor γ, so the distance is $3.46c$. In Mary's frame, the time it takes to reach the star is the distance $3.46c$, divided by the velocity, $0.866c$—equal to 4 years. So in both the Earth frame and in Mary's outgoing frame, she is 4 years old when she reaches the star. Likewise, she will age another 4 years on the return journey and will be 8 years old when she returns.

In the Earth frame, John is at rest. Mary's trip takes 8 years out and 8 years back. When Mary returns, John will be 16 years older.

Now let's examine the same events from Mary's proper frame. This is an accelerating frame, so we do the calculation in three stages. First we use her outgoing frame, moving with velocity $+v$ with respect to the Earth frame. Then she comes to rest on the distant planet; her proper frame becomes identical to John's frame; finally, she accelerates back and her proper frame is one moving at velocity $-v$ with respect to the Earth frame.

The results are shown in Figure A.1. In the first stage, from Earth to the star, in her proper frame Mary is at rest. John is moving at $-v$ and is aging at rate $1/\gamma$. It takes Mary 4 years to reach the star (of course, in that frame, it is the star that is reaching her; she is at rest). In that time, John ages only $4/\gamma = 2$ years.

Then Mary stops at the star (presumably on a nearby planet, not the star itself). Now her proper frame is identical to that of the Earth, so although she is 4 years older, John (in this frame) is simultaneously age 8. This is the first time jump. John didn't suddenly age; it is that Mary has changed Lorentz frames, and in her new proper frame, events that were simultaneous in the old frame are no longer simultaneous in the new frame. Mary knows that in the outgoing frame (in which she is no longer at rest), John is still younger than she is. But in the planet frame, identical to the Earth frame, John is older. John and Mary would both agree to these facts.

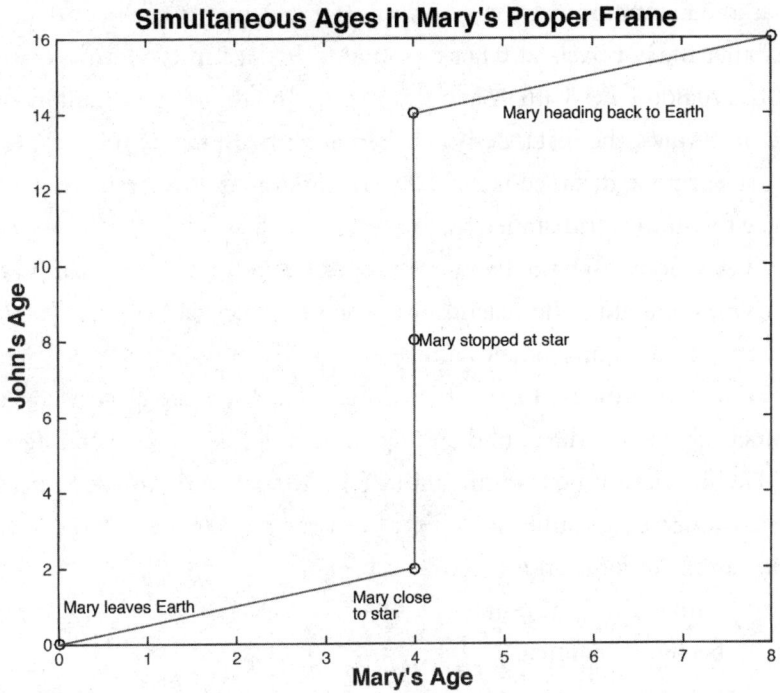

Figure A.1. The twin paradox, showing the age of John in Mary's accelerating proper frame. John's age jumps when Mary's frame changes velocity.

Note from the diagram that the "jump" in John's simultaneous age was 6 years (from 2 to 8). That relates to the time jump equation given earlier:

$$\Delta t = \gamma(\Delta T - \Delta X v/c^2)$$

Here Δt is the jump in John's age. (His age in the Earth frame is equal to the time in the Earth frame.)

Now Mary makes a second change in her proper frame; she accelerates to return. We put in $\Delta X = -3.46c$ (the distance in the return frame), $\Delta T = 0$ (the events are simultaneous), $\gamma = 2$, and $v/c = -0.866$ to get

$$\Delta t = 2(0 + 3.46 \times 0.866) = 6 \text{ years}$$

That is the second jump in John's age, comparing his age in the frame before Mary got back on the spaceship, to his age in the return frame, both simultaneous with Mary's fourth birthday. John's simultaneous age, in Mary's accelerated proper frame, goes from 8 to 14. As Mary returns, John ages an additional 2 years, and he is 16 years older when Mary finally gets back to Earth.

Thus, calculated in both John's proper frame (not accelerating) and Mary's proper frame (accelerating), when they get back together, John has aged 16 years, and Mary has aged 8.

In general, you never want to do a calculation using accelerating frames, if you can avoid it. The jumps in simultaneity are so counterintuitive that they are tricky to handle. Simply stick with any nonaccelerating frame and trust that you'll get the same answer even if you do the calculation the hard way.

Math of the Tachyon Murder

Let's make event 1 the firing of the tachyon gun, and event 2 the death of the victim. $\Delta t = t_2 - t_1 = +10$ nanoseconds, and $\Delta x = x_2 - x_1 = 40$ feet. That means the tachyon is moving at $40/10 = 4$ feet per nanosecond, about $4c$. The plus sign means that the victim dies *after* I fire my gun, since the time value of the death is greater than that of the trigger pull.

Now let's consider the two events in a frame moving at $v = \frac{1}{2}c$. So $\beta = 0.5$, $\gamma = 1/\sqrt{(1 - \beta^2)} = 1.55$. We use the time jump equation:

$$\Delta T = \gamma(\Delta t - \Delta x v/c^2)$$
$$= \gamma \Delta t [1 - (\Delta x/\Delta t)(v/c^2)]$$

Plugging in $\gamma = 1.55$, $\Delta t = 10$ nanoseconds, $v/c = 0.5$, and $\Delta x/\Delta t = 4c$, and canceling the factors of c, gives

$$\Delta T = (1.55)(10 \text{ nanoseconds})[1 - (0.5)(4)]$$
$$= -15.5 \text{ nanoseconds}$$

The fact that the time interval is *negative* means that the order of events is reversed. The victim is shot at time T_2, but since $T_2 - T_1$ is less than zero, T_1 is a bigger number. Therefore T_1, the firing of the gun, took place at a larger—that is, later—time.

Note also that if $\Delta x/\Delta t = V_E$ is less than the speed of light c—that is, if the bullet is subluminal in its velocity—this reversal is not possible. For the reversal, V_E/c must be greater than c/v, and c/v is always greater than 1. So, for any two events that can be connected by a signal going slower than the speed of light, the order in which they occur is the same for all valid frames—that is, for all frames in which v is less than c. We call such events *time-like*. Space-like events are those so far apart that the speed of light is insufficient to connect them.

Math of the Gravity Time Effect

Einstein postulated that the time behavior in a gravitational field could be calculated by assuming it was equivalent to that of an accelerated reference frame. That's what we'll do here.

Suppose we have a rocket of height h, and it is out in a region of space that has no gravity. The rocket is accelerating upward at the Earth's gravity rate, $g = 32$ feet per second every second. Let's assume that the top and bottom of the rocket are accelerated simultaneously, in the frame of the original rocket. After time Δt, its proper frame is moving at velocity $v = g\Delta t$ with respect to its prior proper frame.

We use the equation from the tachyon murder to calculate the corresponding time interval at the top of the rocket:

$$\Delta T = \gamma(\Delta t - \Delta x v/c^2)$$

Putting in $\Delta x = h$ and $v = g\Delta t$, and making the approximation (for nonrelativistic velocities) that $\gamma = 1$ yields

$$\Delta T = \Delta t - hg\Delta t/c^2$$

Dividing by Δt gives

$$\Delta T / \Delta t = 1 - gh/c^2$$

This shows that at height h, the time interval for the top, ΔT, is less than the time interval at the bottom, Δt. Clocks at high altitude run faster.

More generally, the equation is often written as

$$\Delta T / \Delta t = 1 - \emptyset/c^2$$

where $\emptyset$ is the gravitational potential difference. For example, the potential on the surface of the Earth, compared to infinity, is $\emptyset = GM/R$, where M is the mass of the Earth, and R is its radius.

Many textbooks derive the formula using a totally different approach, looking at the redshift of light traveling from the top of a box to the bottom. I prefer the approach I just described because it explicitly uses the equivalence principle that formed the basis of Einstein's general relativity, and it shows that the effect comes from the xv/c^2 term in the Lorentz equations, the same term that leads to the loss of simultaneity.

APPENDIX 2.
TIME AND ENERGY*

The most fascinating, precise, and (for the physicist) practical definition of energy is the most abstract one—too abstract even to be discussed in the first few years of a college physics education. It is based on the observation that the true equations of physics, such as $E = mc^2$, will be as true tomorrow as they are today. That's a hypothesis that most people take for granted, although some people continue to test it; if a deviation is found, it will be one of the most profound discoveries in the history of science.

In the jargon of physics, the fact that the equations don't change is called *time invariance*. It doesn't mean that things in physics don't change; as an object moves, its position varies with time, its velocity varies with time, lots of things in the physical world change with time—but not the equations that describe that motion. Next year we will still teach that $E = mc^2$, because it will still be true.

Time invariance sounds trivial, but expressing it mathematically can lead to an astonishing conclusion: proof that energy is conserved. The proof was discovered by Emmy Noether. Like Einstein, she fled Nazi Germany and came to live in the United States.

Following the procedure outlined by Noether, starting with the equations of physics, we can always find a combination of variables (position,

---

* Adapted from *Energy for Future Presidents* (W. W. Norton, 2012).

speed, and so on) that will *not* change with time. When we apply this method in the simple cases (classical physics, with force and mass and acceleration), the quantity that doesn't change with time turns out to be the sum of kinetic and potential energy—in other words, the classical energy of the system.

Big deal. We already knew that energy is conserved.

But now there is a fascinating philosophical link. There is a reason why energy is conserved; it's because of time invariance!

And there is an even more important result: the procedure works even when we apply the method to the much more complex equations of modern physics. Imagine the following question: In the theory of relativity, what is it that is conserved? Is it energy, or energy plus mass energy? Or something else? And what about chemical energy? And potential energy? How do we calculate the energy of an electric field? What about quantum fields, such as those that hold the nucleus together? Should they be included? Question after question with no intuitive answer.

Today, when such questions arise, physicists apply the procedure outlined by Noether and get the unambiguous answer. Apply the method to Einstein's relativistic equations of motion and you'll get the new energy, one that contains mass energy, mc^2. When we apply the Noether method to quantum physics, we come up with terms that describe the quantum energy.

Does this mean that the "old energy" was not conserved? Yes it does; if we have improved equations, then not only are the predicted motions of particles different, but also the things we thought were conserved are not conserved. Classical energy is no longer constant; we must include the mass energy—and the energy of the quantum fields. By tradition, we call the conserved quantity the "energy" of the system. So although energy itself doesn't change with time, as we dig and uncover the deeper equations of physics, our definition of energy does change with time.

Think about this question: Do the same physics equations that work in New York City also work in Berkeley? Of course. Actually, that observation is not trivial; it has extremely important consequences. We say that the equations don't depend on location. We may have different masses, or different electric currents—but those are the variables. The key question is whether the equations that describe the physics of the behavior of objects and fields is different in different places.

The equations that we have in physics today—all those that are part of the standard physics, the ones that have been verified experimentally— have the property that they work everywhere. Some people think this is amazing enough that they spend their careers looking for exceptions. They look at things that are very far away, such as distant galaxies or quasars, hoping to find that the laws of physics are a little bit different. So far, no such luck.

Now to the remarkable consequence. The same Noether math that worked for equations that don't change with time also works for equations that don't change with location. If we use Noether's method, we can find a combination of the variables (mass, position, velocity, force) that does not change with position. When we apply this procedure to the classical physics invented by Newton, we get a quantity that is equal to the mass multiplied by the velocity; that is, we get the classical momentum. Momentum is conserved, and now we know why. It's because the equations of physics are invariant in space.

The same procedure can be used in the theory of relativity and in quantum physics, and in the combination called *relativistic quantum mechanics*. The combination that doesn't change with time is a little different, but we still call it the momentum. It contains relativistic terms— as well as the electric and magnetic fields, and quantum effects—but by tradition we still call it the momentum.

The close connection between time and energy carries over into quantum physics and its uncertainty principle. According to quantum physics, even though we can define them, the energy and momentum of a part of a system are often uncertain. We may not be able to determine the energy of a particular electron or proton, but the principle does not have a similar uncertainty for the total energy of a system. The whole collection together can shift energy among its various parts, but the total energy is fixed; energy is conserved.

In quantum physics, the time behavior of a wave function has the term e^{iEt}, in which $i = \sqrt{-1}$, E is the energy, and t is the time. When Dirac solved his equation for the electron, he found that it contained negative energies, and that is what caused him to decide that the universe is full of an infinite sea of negative-energy electrons. Feynman found a different interpretation. He suggested that it wasn't E that was negative, but the value of t that appeared in the product. Instead of negative energy, he had electrons moving backward in time. That's what he identified as positrons.

In relativity theory, physicists see space and time as deeply intertwined, and the combination is called *space-time*. The invariance of physics in time leads to energy conservation. The invariance of physics in space leads to momentum conservation. If we put the two together, the invariance of physics in space-time leads to the conservation of a quantity called *energy-momentum*. Physicists see energy and momentum as two aspects of the same thing. From this point of view, physicists will tell you that energy is the fourth component of a four-dimensional energy-momentum vector. If the three components of momentum are labeled p_x, p_y, and p_z, then the energy-momentum vector is (p_x, p_y, p_z, E). Different physicists order the four components differently. Some think energy is so important that they like to put it first. They then call energy the zeroth component of the vector instead of the fourth: (E, p_x, p_y, p_z).

Electric and magnetic fields are also unified by relativity theory, but in a more complicated way. Instead of the 3D vector of the electric field, (E_x, E_y, E_z), and the 3D vector of the magnetic field, usually written as (B_x, B_y, B_z), in relativity theory they become components of a 4D tensor called F, standing for *field*, written like this:

$$F = \begin{bmatrix} 0 & -E_x & -E_y & -E_z \\ E_x & 0 & -B_z & B_y \\ E_y & B_z & 0 & -B_x \\ E_z & -B_y & B_x & 0 \end{bmatrix}$$

That seems rather complicated, with each component appearing twice, but it has the advantage that in a different reference frame, we obtain the new F by applying the same relativity equations we used for position and time. In addition, instead of including electric and magnetic fields separately in our equations, we just include F. That makes the equations look simpler. Doing this *unified* the electric and magnetic fields—that is, made them seem part of one larger object, the field tensor, rather than separate entities.

APPENDIX 3.
PROOF THAT $\sqrt{2}$ IS IRRATIONAL

If we assume that $\sqrt{2}$ is rational—that is, that it can be written as I/J where I and J are both integers, we'll reach a contradiction, and that will prove the assumption false.

If both I and J are even, then we can cancel the common factor of 2 and repeat as necessary until at least one of the integers is odd. That means that if we can write $\sqrt{2} = I/J$, we can also write $\sqrt{2} = M/N$, where at least one of M or N is odd, maybe both.

$M/N = \sqrt{2}$. We square this equation and cross-multiply to get $M^2 = 2N^2$. Since M^2 is a multiple of 2, M^2 is even. That means M is even, since the square of an odd number is always odd. Now I'll show that N is even too.

Since M is even, we can write $M = 2K$, where K is another integer. Squaring this equation gives $M^2 = 4K^2$. We previously showed that $M^2 = 2N^2$, so $2N^2 = 4K^2$. We divide by 2 to get $N^2 = 2K^2$. That means N^2 is even, which means N is even too.

We have contradicted our result that at least one of the numbers M or N must be odd. The only possible cause (since otherwise we followed the rules of mathematics) is that our original assumption—that $\sqrt{2}$ can be written as I/J—is false. Thus the irrationality of $\sqrt{2}$ is proved.

What makes this result so fascinating is that it never could have been discovered through the science of physics. No measurement could demonstrate that $\sqrt{2}$ is irrational. The fact that $\sqrt{2}$ is irrational is a truth that is beyond physical measurement; it exists only in the minds of humans. It is nonphysics knowledge.

If you are interested, you might now enjoy trying to use the identical method to prove that $\sqrt{4}$ is irrational. Of course, it isn't; $\sqrt{4} = 2/1$. Try applying the approach we used here and see where it fails.

APPENDIX 4.
THE CREATION*

At first
there was nothing
no Earth, no Sun
no space, no time
nothing

Time began
and the vacuum exploded, erupted
from nothing, filled with fire
everywhere
furiously hot and bright.

Fast as light, space grew,
and the firestorm grew
weaker. Crystals appeared
droplets
of the very first matter. Strange matter
fragile bits
a billionth of the universe

---

* Previously published in *Physics and Technology for Future Presidents* (Princeton University Press, 2010).

overwhelmed in turbulence
of no importance
they seem
as they wait
for the violence to subside.

The universe cooled and the crystals shattered
and shattered again,
and again and again
until they could shatter no more. Fragments
electrons, gluons, quarks,
grasped at each other, but were burned back apart
by the blue-white heat, still far too hot
for atoms to endure.

Space grew, and the fire diminished
to white to red to infrared
to darkness.
A million year holocaust had passed.
Particles huddle in the cold and bind themselves
into atoms—hydrogen, helium, simple atoms
from which all else is made.

Drawn by gravity, the atoms gather
and divide
and form clouds of all sizes
stars and galaxies
of stars, clusters of galaxies. In the voids
there is empty space
for the first time.

In a small star cloud, a clump of cool matter
compresses and heats

and ignites
and once again there is light.

Deep within a star, nuclei
are fuel and food, burning and cooking
for billions of years, fusing
to carbon and oxygen and iron, matter of life
and intelligence, born slowly, buried
trapped
deep within a star.

Burned and burdened, a giant star's heart
collapses. Convulses. A flash. In seconds
energy from gravity, thrown out
overheats, explodes, ejects
the shell of the star. Supernova! Growing brighter
than a thousand stars. Still brighter, brighter
than a million stars, a billion stars, brighter
than a galaxy of stars. Cinders of carbon, oxygen, iron
expelled into space
escape
free! They cool and harden
to dust, the ashes of a star
the substance of life.

In Milky Way galaxy at the edge of Virgo Cluster
(named five billion years later, for a mother),
the dust divides and gathers and begins to form
a new star. Nearby a smudge of dust begins to form
a planet. The young sun
compresses, and heats
and ignites
and warms the infant Earth

APPENDIX 5.
THE MATH OF UNCERTAINTY

The uncertainty principle in physics is simply a consequence of the fact that particles have wave properties.

The fundamental mathematics of waves has been long understood, and a famous theorem says that virtually any pulse can be represented as a sum of infinite but regular waves (sines and cosines). The subject is called *Fourier analysis*, and it is part of advanced calculus. A favorite problem assigned to students is to build a square wave (one that looks like a series of boxes) out of a sum of sines and cosines.

Fourier analysis has a very important theorem: If a wave consists only of a short pulse, such that most of it is located in a small region Δx (read that as "delta x"), then to describe it in terms of sines and cosines will take many different wavelengths. The wavelengths, in math, are commonly described by a number k. This number is such that $k/2\pi$ is the number of full waves (full cycles) that fit into 1 meter. Physicists call k the spatial frequency or *wave number*. A wave confined to a region Δx in size must contain a range of different spatial frequencies, Δk. Then the Fourier math theorem says that these two ranges have the following relationship:

$$\Delta x \, \Delta k \geq \tfrac{1}{2}$$

This equation has nothing to do with quantum behavior; it is a result of calculus. This theorem predated Heisenberg; Jean-Baptiste Joseph

Fourier died in 1830. It's just math—the math of waves, water waves, sound waves, light waves, earthquake waves, waves along ropes and piano wires, waves in plasmas and crystals. It is true for all of them.

In quantum physics, the momentum of a wave is Planck's constant h divided by the wavelength. The wavelength is $2\pi/k$. That means we can write the momentum (traditionally designated by the letter p) as $p = (h/2\pi)k$. Taking differences for two values of p, this equation becomes $\Delta p = (h/2\pi)\Delta k$. If we multiply the Fourier analysis equation $\Delta x\,\Delta k \geq \frac{1}{2}$ by $h/2\pi$, we get

$$(h/2\pi)\,\Delta x\,\Delta k \geq \tfrac{1}{2}\,(h/2\pi)$$

Then we substitute $\Delta p = (h/2\pi)\,\Delta k$ to get

$$\Delta x\,\Delta p \geq h/4\pi$$

This is Heisenberg's famous uncertainty principle. That's why I said that once we accept that all particles move like waves, the uncertainty principle is a mathematical consequence.

In math, the theorem wasn't really an uncertainty principle; rather it described the range of spatial frequencies in a short pulse. But in quantum physics, the range of frequencies translates into an uncertainty of momentum; the width of the pulse becomes an uncertainty of where the particle will be detected. That's because of the Copenhagen probability interpretation of the wave function. If different momenta (velocities) and different positions are available in the wave function, then making a measurement (such as observing it being deflected in a magnetic field) means picking one out, choosing one value out of many. As Forrest Gump said about life, "[It's] like a box of chocolates. You never know what you're gonna get."

APPENDIX 6.
PHYSICS AND GOD

Physics is not a religion. If it were, we'd have a much easier time raising money.

— *Leon Lederman (discoverer of the muon neutrino)*

Physicalism is the denial of all reality that cannot be measured. Many physicists accept physicalism as the basis of their research but continue to accept the spiritual world as an important, if not the most important, part of reality and of their lives. Some people have the misimpression that all physicists are atheists, and it is worthwhile dispelling that notion. It is completely legitimate for a scientist to dispute a religion that practices science, whether it is a statement by a church that the universe is only four thousand years old, or a claim that Darwinian evolution did not occur. But likewise, there is legitimate criticism against atheists/physicalists who argue that logic and reason are sufficient to deny the existence of a spiritual reality.

I'll continue with a sampling of what some great scientists have said on this subject. Much of this list was compiled with the help of the free e-book *50 Nobel Laureates and Other Great Scientists Who Believe in God*, compiled by Tihomir Dimitrov and available at http://nobelists.net. The references for many of these quotes can be found there.

CHARLES TOWNES (he had everyone, including his graduate students, call him "Charlie"), inventor of the laser and maser, a fellow professor at Berkeley and a personal friend, told me he thought that atheism was "silly." He felt it was denying the "obvious" existence of God. He is quoted (in the book *Science Finds God*, by Sharon Begley) as saying,

> I strongly believe in the existence of God, based on intuition, observations, logic, and also scientific knowledge.

Note that Townes did not include "faith" in this list. When you see something, then acknowledging its existence doesn't require faith. He wrote,

> As a religious person, I strongly sense the presence and actions of a creative Being far beyond myself and yet always personal and close by. . . .
>
> In fact, it seems to me, a revelation can be viewed as a sudden discovery of understanding of man and man's relation to his universe, to God, and his relation to other men.

ARNO PENZIAS, codiscoverer of the cosmic microwave radiation—the radiation that confirmed the Big Bang theory—wrote,

> God reveals Himself in all there is. All reality, to a greater or lesser extent, reveals the purpose of God. There is some connection to the purpose and order of the world in all aspects of human experience.

ISIDOR ISAAC RABI, discoverer of nuclear magnetic resonance (used for MRI, magnetic resonance imaging) and chair of the Atomic Energy Commission, wrote in *Physics Today*,

> Physics filled me with awe, put me in touch with a sense of original causes. Physics brought me closer to God. That feeling stayed with

me throughout my years in science. Whenever one of my students came to me with a scientific project, I asked only one question, "Will it bring you nearer to God?"

ANTHONY HEWISH, codiscoverer of pulsars, wrote in 2002,

> I think both science and religion are necessary to understand our relation to the Universe. In principle, science tells us how everything works, although there are many unsolved problems and I guess there always will be. But science raises questions that it can never answer. Why did the big bang eventually lead to conscious beings who question the purpose of life and the existence of the Universe? This is where religion is necessary. . . .
>
> Religion has a most important role in pointing out that there is more to life than selfish materialism.
>
> You've got to have something other than just scientific laws. More science is not going to answer all the questions that we ask.

JOE TAYLOR, who won his Nobel Prize for discovering rapidly rotating stars that, it turned out, were emitting gravitational waves, wrote,

> We believe that there is something of God in every person and therefore human life is sacrosanct and one needs to look for the depth of spiritual presence in others, even in others with whom you disagree.

Physicalism can be a religion, but it can also simply define the working parameters of physics research, and not be taken to cover all of reality.

Me

Other scientists, writing books like this one, feel it appropriate to describe their own spiritual beliefs. So perhaps it is appropriate for me to make some brief remarks about mine. I hesitate to call these *beliefs*. People believe in the tooth fairy, in Santa Claus, in physicalism. I regard what I am about to say as knowledge based on observation—aphysical spiritual observation, but still observation.

You might call me an aphysicalist. It is not logical to deny observations just because they cannot be measured. I think I have free will, but I recognize that much of it could be an illusion. When I grow hungry, my instincts lead me to seek food, and that is not part of free will. But I do know I have a soul, something that goes beyond consciousness, and that makes me hesitant to allow Scotty to beam me up. I do pray every day, although I am not sure to whom. A wise friend, Alan Jones, once suggested to me that there are only three legitimate prayers: *Wow!*, *Thank you!*, and *Help!* I'm not sure I understand the difference between *Wow!* and *Thank you!* The *Help!* prayer is a request for spiritual, not material, strength. My daily prayers so far have been completely dominated by the words *Thank you!*

Why did we have a Big Bang in the first place? Some have invoked the anthropic principle; others invoke God. I see no good answer. If it was God, that doesn't answer the question of whether God the creator is worthy of being worshipped. Do we venerate a Supreme Being just because he set up some physics equations and lit a fuse? Not I. When I worship, I worship the God who cares about me and gives me spiritual strength.

The ancient Gnostics felt the same way. They believed in two gods: Yahweh the creator, and the other god, the God of Knowledge of Good and Evil. They worshipped only the second. They believed that Adam and Eve did the same. The eating of the apple, in the Gnostic interpretation, was a heroic act. Adam and Eve paid for this "sin" by their banishment from Eden, but they never looked back. To Adam and Eve, nonphysics knowledge was far more important than free fruit.

ACKNOWLEDGMENTS

I am grateful to many people who looked at drafts of this book and suggested improvements and new ideas. These include Jonathan Katz, Marcos Underwood, Bob Rader, Dan Ford, Darrell Long, Jonathan Levine, Andrew Sobel, and members of my immediate family: Rosemary, Elizabeth, Melinda, and Virginia.

My editor, Jack Repcheck, once again offered wonderful guidance and significant assistance in making this book a meaningful whole. He tragically died just as the book was nearing fruition. I thank John Brockman for important suggestions about the tone and style, and for his help in turning an idea into a publishable book. Stephanie Hiebert performed a miracle with the copy editing, and I am very grateful to Lindsey Osteen for her tireless effort tracking down image permissions.

I've benefited by recent discussions of the physics of time and entropy with Shaun Maguire, Robert Rohde, Holger Muller, Marv Cohen, Dima Budker, Jonathan Katz, Jim Peebles, Frank Wilczek, Steve Weinberg, Paul Steinhart, and many other colleagues and friends.

CREDITS

Part I cover image: Albrecht Dürer/Wikimedia

Figure 2.1: Lucien Chavan/Wikimedia

Figure 2.2: Personal photo provided by Richard A. Muller

Chapter 3 epigraph: "As Time Goes By," written by Herman Hupfeld. Used by permission of Bienstock Publishing Company o/b/o Redwood Music Ltd.

Figure 3.1: Unknown photographer/Wikimedia

Figure 4.1: Joey Manfre

Figure 5.1: © 2014 The Regents of the University of California, through the Lawrence Berkeley National Laboratory

Figure 6.1: "Einstein 1921" by F. Schmutzer. Restored by Adam Cuerden.

Figure 6.2: *Calvin and Hobbes* © 1992 Watterson. Reprinted with permission of UNIVERSAL UCLICK. All rights reserved.

Figure 7.1: Richard A. Muller

Figure 7.2: Richard A. Muller

Part II cover image: © Hayati Kayhan/Shutterstock

Figure 9.1: Bjoern Schwarz

Figure 10.1: Grave of Ludwig Boltzmann, physicist, Zentralfriedhof (Central Cemetery), Vienna, Austria. Photograph taken by Daderot [http://en.wikipedia .org/wiki/User:Daderot], May 2005.

Figure 11.1: Bain News Service/Library of Congress

Figure 12.1: NASA/JPL-Caltech

Figure 12.2: Richard A. Muller

Figure 12.3: Wikimedia

Figure 12.4: *Calvin and Hobbes* © 1989 Watterson. Reprinted with permission of UNIVERSAL UCLICK. All rights reserved.

Figure 13.1: NASA; ESA; G. Illingworth, D. Magee, and P. Oesch, University of California, Santa Cruz; R. Bouwens, Leiden University; and the HUDF09 Team

Figure 13.2: US Department of Energy

Figure 13.3: Richard A. Muller

Figure 13.4: NASA/WMAP Science Team

Figure 14.1: US Department of Energy

Figure 14.2: Richard A. Muller

Figure 15.1: US Department of Energy

Figure 15.2: *Calvin and Hobbes* © 1990 Watterson. Reprinted with permission of UNIVERSAL UCLICK. All rights reserved.

Figure 16.1: Catwalker / Shutterstock.com

Part III cover image: NASA/ESA/Hubble Heritage Team (STScI/AURA)/J. Hester, P. Scowen (Arizona State U.)

Figure 17.1: Christian Schirm

Figure 17.2: Benjamin Schwartz, The New Yorker Collection/The Cartoon Bank

Figure 18.1: Edmont/Wikimedia

Figure 19.1: Richard A. Muller

Figure 19.2: Richard A. Muller

Figure 19.3: James Clerk Maxwell

Figure 20.1: Carl David Anderson; modified by Richard A. Muller

Figure 20.2: Nobel Foundation/Wikimedia

Figure 20.3: Richard A. Muller

Figure 20.4: Richard A. Muller

Figure 20.5: Richard A. Muller

Part IV cover image: "Rotating Rings," © Gianni A. Sarcone, www.giannisarcone.com <http://www.giannisarcone.com>. All rights reserved.

Part V cover image: © OGphoto/iStock.com

Figure 24.1: *Zits* used with the permission of the Zits Partnership, King Features Syndicate and the Cartoonist Group. All rights reserved.

Figure A.1: Richard A. Muller

INDEX

Page numbers followed by *n* refer to footnotes. Page numbers in *italics* refer to figures and illustrations.